TECHNOLOGY ON TRIAL

PUBLIC PARTICIPATION
IN DECISION-MAKING
RELATED TO
SCIENCE AND TECHNOLOGY

ORGANISATION FOR ECONOMIC CO-OPERATION AND DEVELOPMENT
PARIS 1979

The Organisation for Economic Co-operation and Development (OECD) was set up under a Convention signed in Paris on 14th December 1960, which provides that the OECD shall promote policies designed:
- to achieve the highest sustainable economic growth and employment and a rising standard of living in Member countries, while maintaining financial stability, and thus to contribute to the development of the world economy;
- to contribute to sound economic expansion in Member as well as non-member countries in the process of economic development;
- to contribute to the expansion of world trade on a multilateral, non-discriminatory basis in accordance with international obligations.

The Members of OECD are Australia, Austria, Belgium, Canada, Denmark, Finland, France, the Federal Republic of Germany, Greece, Iceland, Ireland, Italy, Japan, Luxembourg, the Netherlands, New Zealand, Norway, Portugal, Spain, Sweden, Switzerland, Turkey, the United Kingdom and the United States.

***

© OECD, 1979
Queries concerning permissions or translation rights should be addressed to:
Director of Information, OECD
2, rue André-Pascal, 75775 PARIS CEDEX 16, France.

TABLE OF CONTENTS

Preamble	5
Foreword	7
Introduction	11

I. SCIENCE, TECHNOLOGY AND PUBLIC PARTICIPATION
- A. Forms of Participation 15
- B. Scientific and Technological Issues 16
- C. Framework for Analysis 19

II. INFORMING THE PUBLIC
- A. Access to Information 21
- B. Information on Decision-Making 24
- C. Public Understanding 26

III. INFORMING POLICY-MAKERS
- A. Advisory Bodies 56
- B. Legislative Hearings 58
- C. Commissions of Inquiry 61
- D. Special Ad Hoc Mechanisms 77

IV. RECONCILING CONFLICTING INTERESTS
- A. Administrative Decision-Making 82
- B. Regulatory Decision-Making 86
- C. Administrative and Judicial Appeal 91

V. COLLABORATIVE DECISION-MAKING
- A. Science Courts and Citizen Review Boards 97
- B. New Mediation Procedures 101
- C. National, State and Local Referenda 102

VI. CONCLUSION
- A. The Nature of Participatory Phenomena 110
- B. Information and Public Understanding 111
- C. Participation in Decision-Making 113

ANNEXES

I. List of Participants 115
II. References 117

PREAMBLE

The work of the Committee for Scientific and Technological Policy with respect to public participation has been undertaken in response to a recommendation of its Meeting at Ministerial level in June, 1975, when Ministers suggested that OECD examine Member countries' experience regarding public participation in government decision-making related to science and technology. It was hoped that this initiative would assist them "to devise effective means of informing the public of the implications of new technological developments, soliciting their reactions, and engaging them in the decision-making process".

This activity has comprised a comparative study of the experiences of three Member countries—Austria, the Netherlands and Sweden—undertaken in 1976.[1] Drawing on this preliminary analysis, additional detailed case studies were carried out in 1977 on other national experiences with public participation in decision-making related to science and technology. The data and findings from all these analyses served as a basis for discussions by a group of specialists in late 1976 and early 1978.[2]

This Report has been written by K. Guild Nichols, a member of the Science Policy Division of the OECD Directorate for Science, Technology and Industry. It served as a background document for a "forum discussion" held by the OECD Committee for Scientific and Technological Policy in October, 1978. Its findings have been updated as of the spring of 1979.

It goes without saying that the Report does not attempt to provide any general and definitive conclusions on such a delicate and fluctuating topic where specific national traditions, structures and processes often play an essential role. More modestly, its aim is to present a preliminary assessment of the nature of different national experiences, provide a clearer understanding of these participatory phenomena, and identify the various problems and areas of present concern as a guide for future action.

---

1. Nelkin, D., *Technological Decisions and Democracy*, Sage Publications Ltd., London, 1977.
2. See Annex I for a list of persons who participated in these two OECD meetings.

FOREWORD

This Report constitutes a summary of findings based upon an examination of different national participatory mechanisms and experiences. Because the description and analysis are felt to be of general interest and relevance to many other areas of contemporary policy concern, the Committee considers that its publication would provide an important contribution to broader public discussion and debate.

Public participation is a concept in search of a definition. Because it means different things to different people, agreement on what constitutes "the public" and what delimits "participation" is difficult to achieve. The public is not of course homogeneous; it is comprised of many heterogeneous elements, interests and preoccupations. The emergence over the last several decades of new and sometimes vocal special interest groups, each with its own set of competing claims and demands, attests to the inherent difficulty of achieving social and political consensus on policy goals and programmes purporting to serve the common interest.

Public participation in government decision-making therefore takes many different direct and indirect forms. This reflects the diversity of national and political and cultural traditions, structures and processes. That citizens in many countries appear today no longer content to rely solely upon traditional, indirect democratic procedures, but are demanding more direct forms of representation and participation in decision-making, poses special problems for government and new challenges to the viability and adaptability of representative democracy itself.

This so-called "participatory phenomenon" touches many areas of public policy concern. Issues related to science and technology are not unique in this respect. However, as a Committee, we are especially mindful of the critical role that scientific research and technological development does and can play, not just with respect to national economic productivity and social well-being, but to the shaping and enhanced potential of all aspects of society, in developing and developed nations alike. Informed and responsible public participation can and should play an increasingly important and effective role in the articulation of social and political goals and in the elaboration of scientific and technologically-related programmes.

At the same time, one cannot minimize the difficulties of trying to promote increased public participation. Scientific and technologically-related issues do pose a number of special problems. These relate not only to their novelty and complexity and to the difficulty the general public often has in grasping their significance and understanding their technical content, but also to the difficulties of adapting traditional patterns of decision-making on scientific and technological matters to meet public demand for more direct participatory involvement.

The Report presents a general overview of the diverse approaches and mechanisms which some OECD Member countries have devised to respond to these new demands and needs. In addition, it attempts to identify some of the key factors and motivations behind the emergence of these participatory phenomena, and to bring to light specific problems raised by public participation in issues where scientific and technological aspects play a dominant role.

Because of the importance attached to this subject, the Committee for Scientific and Technological Policy held a full-day "forum discussion" on the issue of public participation in government decision-making related to science and technology on 10th October, 1978. In addition to a number of specialists from Member countries, the Committee also invited four guest speakers, representing different professional competences and perspectives, to participate in this discussion: the Honourable Mrs. Kerstin Anér, Swedish Under-Secretary of State for Education and Cultural Affairs; the Honourable Mr. Justice Thomas R. Berger, member of the Supreme Court of British Columbia, Canada; Professor Heinz Maier-Leibnitz, President of the Deutsche Forschungsgemeinschaft; and Dr. Alexander J. Morin, Director of the Office of Science and Society of the US National Science Foundation. We are especially indebted to them for their valuable insights and important contributions to this discussion.

As might be expected, this debate was wide-ranging, covering not only issues and processes related to decision-making on science and technology but also, more generally, to how governments can become more responsive to citizen needs and desires for more direct involvement in shaping the future of their societies. I personally found this "forum discussion" exceedingly valuable and enlightening; both in terms of sharing common experience and concerns and, equally important, in clarifying important divergencies in views and attitudes towards the desirability of encouraging increased public participation in government decision-making. For, after all, public participation is a topic that seldom elicits a neutral response; identifying areas of fundamental disagreement is as important to future policy debate and discussion on this general matter as are the areas of common accord.

Several issues raised in this Report were underscored during the debate and seem to me to merit special mention. First, scientific and technologically-related policy disputes often bring into question certain deeply-held social values. This is especially evident within the area of energy policy. The articulation of scientific research priorities also raises important ethical concerns such as the qualitative aspects of scientific and technological advance, and the price exacted from different sectors of society for economic growth. Whilst greater involvement on the part of the public will not itself, of course, necessarily lead to a consensus on the goals to be pursued, it can provide a very important means of clarifying attitudes and crystallizing opinions in seeking to provide the greatest possible benefit to the largest number of people.

Perhaps the most crucial question raised throughout the discussion concerned the definition of public policy issues themselves. It was argued forcibly that, particularly in areas associated with science and technology, there has been a marked tendency to cloak public policy issues in narrow technical terms, thus obscuring the essentially political nature of choices to be made. As a result, the argument runs, political choices become submerged or confused in debates over technical feasibility. No matter how complex the task, it is imperative that serious efforts be made to disentangle, and hence clarify, the diverse ends and means which become intertwined within contemporary public policy issues. If the creation of a greater sense of shared responsibility is an aim of current attempts to promote public participation—and this seems to me to be an essential aim—then ways must be found to open up public discussions when the choice of policy goals is still open and before the debate on technological means is fully launched.

Efforts to promote more informed public debate by providing greater public access to information and by improving levels of technical understanding have received considerable attention in many Member countries. However, as underlined by some participants at the "forum discussion", whilst a certain level of technical understanding and appreciation on the part of the general public is essential, it is equally, if not more important, that the information provided be politically relevant to the issues to be decided and the value choices to be made. It was further underlined that responsibilities for the provision of such information are shared by government, educational institutions, and other public and private institutions.

Although common problems and concerns manifest themselves in different settings, special attention needs to be given as to whether and how other national approaches and mechanisms can be adapted to serve the special needs and requirements of each national political system. Because of cultural, political and structural differences, it would of course be naive and even dangerous to imagine that specific approaches and measures devised to encourage public participation in one national setting could or should be transferred, *in toto,* to another.

Finally, I find it entirely understandable that increased public participation is viewed by many policy-makers as especially time-consuming, and therefore potentially counter-productive in terms of decision-making efficiency. Nonetheless, as emphasized by a number of delegates, such concerns need to be tempered by the realisation that "timely" programme decisions, in the absence of a sufficiently broad consensus on overall policy goals, can also result in long-term economic and social costs, especially when moving from programme definition to programme implementation.

James MULLIN
Chairman of the Committee for Scientific and Technological Policy

Whether or not a societal organisation is experienced as tyranny seems to depend mainly on whether its members are assured relative free choices and a part in decision-making about issues that for them are the sum of their consciousness of freedom.

Bruno Bettelheim
The Informed Heart

O.K., then, it's settled. We present it to the public with all its pros and cons, we let the media chew on it for a while, we go through a lot of soul-searching, and then we go ahead and do it.

Satirical Cartoon in
The New Yorker

Quis custodiet ipsos custodes?

Decimus Junius Juvenalis
Satires VI, 347.

INTRODUCTION

Scientific discoveries and technological developments have always involved social adjustments and accommodation to new ideas, and the emergence of new values. But, the speed at which the impact of new knowledge and its implications are felt is a characteristic of our present-day society.

What is particularly significant today is the extent to which ordinary citizens appreciate and are concerned by the possible results of scientific and technological activities which were previously considered to be within the province of only a few experts or administrators, who were supposed to know how to deal with them. For example, plans for the siting of new airports, highways, industrial facilities and electricity generating plants have met with increasing public hostility from people concerned about their local impacts. Debates over dangers and risks associated with undertaking certain types of advanced biomedical research or with large-scale technological projects such as gas pipelines, uranium mining, or nuclear waste disposal have developed into major national policy issues. And, national controversies over, for example, the safety, security and desirability of nuclear fuel reprocessing, have become topics of growing international concern.

Because science and technology affect all aspects of contemporary life, none of their effects is immune from public controversy. Nor is the general public alone in raising some of these concerns. Scientists have been among the first to voice reservations, for example, about the uncontrolled pursuit of recombinant DNA research. Increasing numbers of scientists are concerned about not just the ends which scientific research should serve, but the means by which it is carried out, and whether in fact certain types of research ought to be undertaken at all.

Moreover, issues raised in one national context spread rapidly to others. Debates over the use of medical technologies to prolong life, and policies of passive euthanasia to terminate it, are of increasing concern throughout much of the Western world. Fears as to possible misuse of new research techniques for the conduct of large-scale social experiments, or for modifying the weather, are expressed in developing and developed countries alike.

Public preoccupation with the impacts of technological developments, fears about their possible dangers and risks, ethical considerations over the potential misuse and side-effects of scientific research techniques or results are major public concerns behind many contemporary scientific and technological controversies. There is also another central factor: concern about social values. It is this preoccupation with current values—those that some people want to preserve, those that others wish to modify—that lies at the heart of many so-called scientific or technological debates.

One of the key manifestations of this concern can be seen in the increasing demand for greater public participation in government decision-making on scientific and technologically-related issues. Touched by anxiety over the future, by a sense of powerlessness over the present, citizens are seeking more direct ways of influencing the outcome of decisions on matters which they perceive as affecting their lives. In some cases, their motives are altruistic—they purport to seek a more equitable distribution of economic and social costs and benefits. For others, public participation is seen as a means of making governmental agencies more accountable to the people. For most people, however, it is far more personal. They wish to regain a measure of control over their lives—a control they feel has been abrogated by bureaucratic processes which appear to them to be, more often than not, directed to the resolution of technical problems, rather than to meeting human needs.

Individuals and groups are, therefore, seeking greater access to traditional forums of debate and to information in the possession of governmental authorities. Frustrated by seemingly arbitrary decisions, they are making increasing recourse to the courts. People are coming forward in new combinations, often by-passing existing social and political structures, to advocate new positions and alternative strategies in diverse areas of government policy. Consumer advocates, environmental and physical planning groups, health care specialists, alternative energy activists, and many other specialised interest groups are mobilising citizens, politicising issues, and trying to influence public opinion as well as governmental decisions. In some cases, traditional channels of participation have been rejected in favour of new forms of political action and agitation: demonstrations, plebiscites, and overt opposition.

Not surprisingly, for many government functionaries these participatory phenomena are not only new, but troubling. Complex and costly plans have had to be scrapped, or deferred, because of public opposition. Policies prepared after long and intensive study, negotiation and compromise are suddenly called into question by adverse public reactions. Some governments have, therefore, come to recognise the need to develop more effective mechanisms for involving the public in the formulation and implementation of decisions.

This is no easy task, and it is not surprising that government response to participatory demands has been limited and cautious. After all, the reasons for taking up arms for, or against, a particular scientific or technological development are many and diverse. Are their proponents concerned about the results of certain technologies impinging on a specific area of interest or need; or do they reflect a more pervasive disillusionment with technologies in general and with the goals to which they are applied? Or, are they a reaction against the exercise of governmental authority? Are public controversies involving science and technology any different from those in other domains of public policy? Are they not simply an expression of anxiety on the part of the public at being manipulated?

For many elected representatives and appointed government officials there is an additional set of unknowns that also cause concern and hesitation. For example, what are the possible consequences of expanding public participatory processes? What impact will they have on decision-making efficiency as well as on existing administrative and political structures and processes? Are there practical limits to public participation and what are their implications for the concept of representative democracy?

This study attempts to address some of these questions and how governments have responded to the demand for greater participation on the part of the public in decisions concerning science and technology. It looks at the process of public participation largely from the point of view of government, not from the perspective of citizens or citizen groups. Its scope is therefore limited to a description and analysis of government initiatives. It does not purport to analyse in detail public motivations for participation, the broad range of different citizen initiatives or how citizens themselves tend to view government actions with respect to the promotion, conduct or control of public participatory activities.

Moreover, this study is almost exclusively concerned with public participation in affluent, highly-industrialised societies. The majority of the examples cited concern national policy matters; that is, with several exceptions, issues related to regional or local matters are generally not dealt with. Finally, it should be re-emphasised that the central focus of this study is on public participation in government decision-making related to science and technology, and not to government decision-making in general.

I

SCIENCE, TECHNOLOGY AND PUBLIC PARTICIPATION

A. FORMS OF PARTICIPATION

The concept of public participation is an elusive one. It means many different things to many different people. To some it is a political privilege to be kept informed and to be able to vote for persons representing their own views. To others it is an elemental right, a right to be heard and to have one's say, and be directly involved in the exercise of decision-making processes. And to still others it is an opportunity to provide a living expression of identity with the State. To most, however, it is a mixture of these.

The concept of public participation also has different meanings in different national contexts. In those countries with an Anglo-Saxon heritage, it is often closely associated with other similar concepts, such as participatory democracy, direct democracy and grass-roots democracy. All of these concepts generally refer to efforts on the part of people seeking more direct means of influencing decision-making. No longer satisfied with relying solely on the power of their electoral vote, citizens in these countries are increasingly prone to affiliate themselves with specialised interest groups and organisations for the purposes of exerting their collective influence.

Although the exact equivalent of the term "public participation" is not generally used in most Western European countries, other terms are of course employed: "democratisation" and "co-determination", for example. "Co-determination"—or "Mitbestimmung" as it is known in Germany—has its origin in the trade union movement.

In Germany, Austria and the Scandinavian countries, one witnesses the emergence of new "citizen initiative" movements, often coalescing around specific issues or projects. They resemble, in both approach and effect, many of the public participatory movements in the Anglo-Saxon sense of the term.

The definition of public participation employed in this analysis is the following: _public participation is any activity by any person, group of persons or organisation, other than elected or appointed officials of government or public corporations, that directly or indirectly is aimed at taking part in or influencing the affairs, decisions, and policies of the government or public corporation._ The emphasis in this definition is upon direct and indirect forms of public participation by both individuals and groups of individuals. However, by and large, we are explicity leaving out of this definition the role and function of political parties, elections and individual ballot-casting. These latter forms of representation, in combination with the former direct and indirect forms of public participation, are considered to be part of the broader concept of political participation in democratic systems.

We are of course aware of the limitations of such distinctions. However, because we do not envisage to analyse entire political systems, the definition proposed is primarily of analytical validity for the context and purpose of this particular study.

Public participation, as defined for the purposes of this study, includes therefore a broad range of activities: public referenda; membership on governmental or quasi-governmental regulatory boards, advisory councils and committees; interventions before public inquiries and hearings, regulatory agencies, administrative and judicial appeals boards; citizen litigation; lobbying; and public protests and demonstrations. Public participatory activities are not therefore limited only to those that are promoted or sanctioned by government. Neither are they meant to exclusively comprise extra-governmental activities. This definition attempts to bridge these extremes.

Finally, we are primarily concerned with participatory phenomena that are of relatively recent occurrence. That is, we are interested specifically in exploring what is particularly "new" or different with respect to public participation in government decision-making over the last two decades. In most countries, public demand for a more direct role in decision-making on issues related to science and technology is not an isolated phenomenon; it is but one aspect or extension of a more general trend towards increased activism on the part of some citizens in many areas of public policy concern.

B. SCIENTIFIC AND TECHNOLOGICAL ISSUES

Science and technology do, however, have a number of distinguishing characteristics which cause special problems or complications. One of their most obvious overall characteristics is ubiquity: they are everywhere. They are at the forefront of social change. They not only serve as agents of change, but provide the tools for analysing social change. They pose, therefore, special challenges to any society seeking to shape its own future and not just to react to change or to the sometimes undesired effects of change.

There appear to be at least six factors that distinguish scientific and technologically-related issues from other issues commonly associated with public controversies and participatory movements. The *first* of these concerns the rapidity of change brought about through advances in science and technology. The speed with which new scientific ideas and technological innovations are introduced into society often has an unsettling effect on many people. For some, scientific and technological progress, in and of itself, is accepted as a kind of "Manifest Destiny". For others, it is something that must be controlled and directed—if only to avoid potentially dangerous side-effects. But many are unanimous in expressing concern at the speed of change and the general feeling of insecurity it engenders.

Second, is the fact that many such issues are entirely new. Issues involving, for example, genetic engineering, foetal research, weather modification or earthquake prediction are no longer simply the subjects of science fiction speculation but of growing public concern and scientific debate. Because of their novelty and complexity, the general public does not fully understand them nor all of their potential implications. Lack of understanding can result in ambivalence or, more often, in increased fear and uncertainty.

Third, the scale, complexity and interdependency associated with many such issues, especially those involving technological undertakings, are often of an order of magnitude never before encountered. For example, plans for the construction of oil and natural gas pipelines demand an attention to technical detail and a comprehensive treatment of economic, social, cultural, environmental and political implications heretofore seldom associated with many traditional areas of governmental decision-making.

A *fourth* distinguishing characteristic of scientific and technological issues relates to the dimensions and irreversibility of some of their effects. Some are visible, affecting certain interests but not others. Others are less visible and more pervasive; for example, the cumulative impact of nitrogen oxides on the ozone layer or of computerised data banks on personal privacy. While the dimension of still other impacts, such as the effects of exposure to low-dose radiation are often of such a long-term nature that they exceed the present predictive capacity of science.

A *fifth* factor is that many scientific and technological issues raise important ethical concerns or bring into question certain deeply-held social values. Given the pervasiveness of science and technology, almost all people are affected, whether they choose to be or not. The ethical and value dilemmas raised by such issues are therefore of unavoidably general social relevance.

The *sixth* factor relates to public sensibilities about real or imagined threats to human health and perceived dangers inherent in scientific and technological developments. For, despite generally widespread public support for science and technology, there is evident increasingly broad concern over their deleterious side-effects. Present-day preoccupations about toxic chemicals in the environment, dangers in the work-place, or consumer risks in the home are but manifestations of a still more general concern: concern as to the unseen, and unforeseen, dangers and risks to human health of living in highly technologically-oriented societies.

These six factors—the rapid pace of scientific and technological change, the novelty and complexity of some issues and their scale, the dimensions of their impacts, ethical and value concerns and general public perceptions—all of these factors taken together distinguish scientific and technological issues from other areas of public policy concern.

There is, however, another set of important factors. These relate not only to the nature of the issues, as described above, but also to the way in which governmental decision-making functions with respect to scientific and technological matters. Many people feel that there is a tendency within government to define broad political problems in narrow technical terms, and that this inevitably leads to "closed politics": that those lacking sufficient expertise or technical competence are excluded from decision-making forums.

There is also a tendency within government to "parcel out" parts of each issue for special treatment by each competent ministry. One result some people fear is that important overall perspectives are often lost. Another is that everyone and no one appears to be responsible for many of the decisions that are made.

Some also believe that government dependency upon technical expertise and the search for largely-technical "solutions" to many problems associated with science and technology results in political choices being submerged in debates over optimum technical feasibility. The general public is often not only left behind in a state of utter confusion and mystification, but sometimes feels its political choices are pre-empted.

Government secrecy and limited access to many types of technical information are usually justified on the grounds that only those parties having a "special interest" or "competency" with respect to a given issue should be entrusted with the relevant information. Thus, it often becomes difficult for citizens expressing a general interest in an issue or possessing only a limited competency to gain access to all relevant information.

**

Public demand for participation in governmental decision-making on scientific and technological issues can therefore be viewed as originating from two principal sources: from concern about the nature of specific issues, and from disenchantment with contemporary political processes.

Increasingly aware of the scale of technological undertakings and of their potential impacts, people are preoccupied about dangers and risks and the ethical dilemma of who should share them. Limited access to technical forums of debate has roused public suspicion and inspired demands for greater political accountability.

But the general public does not speak with one voice—at least not most of the time. And, for many government administrators and politicians, public demands for participation are often difficult to decipher. Are they the expression of the "public interest" or of the interests of innumerable different publics? Are so-called "public interest" groups representative of the commonweal or, for that matter, even of the views of their own membership?

One implication of the rapid growth of competing public demand for participation is the increasing difficulty encountered in achieving a common consensus about the future. This is all the more evident within the areas of science and technology, where it is extremely difficult to assess long-term policy impacts. Lack of consensus over future goals and policies makes governmental decision-making agonisingly difficult and sometimes impossible. Lack of consensus over how best to achieve consensus often results in acrimonious debate.

How better to inform the general public on technical policy matters is linked to the need for better information on the public's own needs and desires. The setting of government standards for risks is increasingly dependent upon ascertaining levels of social acceptability. And, public demands for greater political accountability are closely tied to the government's need to ensure its own credibility.

Science policy makers have an important contribution to make towards resolving some of these problems. Traditionally concerned with the formulation and articulation of policies for the promotion and orientation of scientific research, training, and technological development, such policy-makers often find themselves not at the centre, but at the periphery of most major public policy debates. Such a position has its advantages, however. Not being a partner to many conflicts with special interests to defend, science policy-makers can benefit from their somewhat detached perspective. Moreover, their relationship to the scientific community also carries special advantages and certain responsibilities. They are in a position to provide an overview of issues and their long-term implications; something which is often lacking when people succumb to "tunnel-vision".

Science policy-makers are thus confronted with a number of challenges. Mindful of the fact that public hostility to science and technology has also had some positive impacts—for example, in terms of improved worker safety and new health regulations—they can seek to achieve a closer harmony between the aims of research and the goals of society.

More specifically, what special roles and responsibilities does the scientific community share with government for promoting broader public understanding, for identifying the boundaries between facts and values, and for distinguishing the limits of scientific knowledge in public policy debates? How can science policy-makers contribute to this and to the development of more active and less reactive approaches to problem-solving, to the need for comprehensive, long-range perspectives, and to the identification of new or emerging areas of scientific and technological controversy?

Responding to the demand and need for public participation is no single or simple responsibility. Moreover, there is every indication that public participation, in science and technology as in other areas, is not a transient phenomenon. It is one that will continue to challenge and preoccupy societies for many years to come.

C. FRAMEWORK FOR ANALYSIS

Government response to public demand for participation in decision-making takes many forms which vary from country to country. Because of differing cultures and political traditions, the "problem" of public participation manifests itself in different ways and is dealt with by different means. Each country has its own special strengths and weaknesses. In Scandinavia, continental Europe and Japan, there is a tradition of a strong chief executive and the maintenance of a strong and effective bureaucracy. The civil service of these countries is comprised of a unified corps of generalists and administrative lawyers, usually closely controlled by and responsive to the dictates of Cabinet. The bureaucracy tends to have a dominating influence on parliamentary proceedings and provides a certain stability and continuity to these political systems, often characterised by fragmentation among numerous political parties.

By contrast, in the more loosely-structured political system of the United States, there is a general demarcation between the career civil service and the department and agency heads who are usually short-term political appointees drawn from outside government; authority is shared between these political appointees and career civil service officers. The United States Chief Executive and his immediate staff do not usually have the same degree of control over the bureaucracy as is exercised by Cabinet in the classic parliamentary systems. Moreover, the United States Congress, with its system of specialised committees and control over budgetary authorisations and appropriations, maintains a close oversight over Executive policies and programmes. It is often able to influence directly policy implementation by controlling the structure and staffing of Executive agencies. At the same time, it does not generally brook incursion by Executive authorities in its legislative proceedings; it tries to protect its own legislative prerogatives by keeping in check Executive power over policy determination.

Traditions of public participation in government and politics are especially strong in the United States, Canada and the United Kingdom. They seem to instill vitality to the democratic process and reinforce popular beliefs that participation is "a good thing". They may cause problems for the governability of those political systems, but these problems have been traditionally considered to be of manageable proportions. In the Scandinavian and other Western European countries and Japan, public participation is sometimes viewed with more ambivalence. Reservations are especially felt in those countries where Executive power is derived from and dependent upon coalitions between several among many political parties. Public participation, especially in the form of large ad hoc citizen initiatives associated with specific national policy issues or disputes, is thus often viewed with concern; not just for the governability of society but for the effective functioning of representative systems of government.

Therefore, government response to demand and need for public participation varies considerably not just between but also within political systems. Appointed executive officers and members of the bureaucracy do not always see eye-to-eye with each other or with elected officials on the aims, methods or conditions necessary for limiting or promoting public participation. This is perhaps not surprising, given the additional fact that the aims of most participants themselves are sometimes unclear, seldom homogeneous, and never static.

Nevertheless, one can identify four broad categories of government response to public pressures for more direct participation in decision-making on issues related to science and technology.[3]

The *first* of these relates to government efforts to inform the general public. A number of new administrative and legal mechanisms have been devised to ensure broader public access to information in the hands of government and to inform citizens on the nature of decision-making processes and on opportunities for their more direct involvement. The promotion of more informed public participation has also served as the major aim of specialised public information campaigns, committees and programmes designed to enhance public understanding on scientific and technologically-related matters.

A *second* category of response has involved the expanded use of existing consultative mechanisms designed to inform policy-makers on the needs, wants and desires of the general public. The use of governmental advisory bodies, legislative hearings, commissions of inquiry and other new ad hoc consultative mechanisms have figured prominently in this regard. In a number of countries, some members of the general public have, nevertheless, manifested their dissatisfaction with these two sets of approaches which are primarily aimed at informing citizens and policy-makers; they have demanded more direct participation in the actual processes of governmental decision-making.

Therefore, a *third* category of response has been to seek to reconcile conflicting interests by providing expanded opportunities for citizen intervention in government administrative and regulatory proceedings. In addition, citizens themselves have taken recourse to legal measures as a participatory device aimed at ensuring that their views are recognised and their interests taken into account.

Finally, attempts have been made to provide opportunities for citizens to express their opinions through the development of collaborative forms of decision-making. The most direct expression of such collaborative participation is the national, state or local referendum. Two additional approaches involve the use of citizen review boards and new mediation procedures.

These four categories are not mutually exclusive. There is some necessary and inevitable overlap between categories of response and between the objectives sought by the various mechanisms employed. The purpose served by this framework is pragmatic and policy-oriented: to provide a means for sorting out different problems associated with government attempts at promoting public participation, and for analysing the salient features that distinguish participation on scientific and technologically-related issues and controversies from other areas of major public policy concern.

---

3. For further applications of this framework see, for example: Vindasius, D., *Public Participation Techniques and Methodologies: A Resumé,* Department of the Environment, Ottawa, 1974; Hampton, W., "Research into Public Participation in Structure Planning", in Sewell, R.D. and Coppock (eds.), *Public Participation in Planning,* Wiley and Sons, London, 1977; and Nelkin D. and Fallows, S., "The Evolution of the Nuclear Debate: The Role of Public Participation" in *Annual Review of Energy* (USA), Vol. 3, 1978, pp. 275-312.

II

INFORMING THE PUBLIC

As the major producer and consumer of information in contemporary society, government has derived considerable power from its enhanced capabilities for managing the collection, interpretation, use and dissemination of information. At the same time, information is now seen as an essential means for individual advancement and a prerequisite for citizen participation in the democratic process. As underscored by one recent government report, "no matter how skilled an individual may be, how sophisticated the techniques at his disposal, or how perceptive his judgment, a lack of adequate information will seriously prejudice him and impair the effectiveness of any decision he has to make".[4]

Public demand for disclosure of information in possession of government has figured prominently in many recent technologically-related policy debates. The central thrust of these demands has been for information that reveals the functioning of the decision-making process and the basis upon which decisions are made. Coupled with these concerns is a growing questioning about government reliance on technical expertise and the ways in which it is employed in decision-making on scientific and technological matters. Cognisant of these growing concerns and of the need for a more informed public participation, governments have undertaken a number of measures aimed at more effectively informing the general public.

These measures can be described in terms of three inter-related objectives: promoting broader public access to government information; increasing knowledge about decision-making procedures and participatory opportunities; and increasing levels of public understanding in areas of science and technology.

A. ACCESS TO INFORMATION

The Scandinavian countries and the United States were the first to adopt national legislation providing for enforceable right of access to government documents.[5] Australia, Canada and the Netherlands have also been considering similar proposals for freedom of information legislation for nearly half a decade, while other countries, such as the United Kingdom, France and West Germany, although not having enacted such legislation, have given consideration to making

---

4. *Report of the Royal Commission on Australian Government Administration (Coombs Report)*, Vol. 1, Section 10.7, AGPS, Canberra, 1976.
5. Freedom of the Press Act (Sweden), The Law on The Public Character of Official Documents (Finland), Law on Publicity in Administration (Denmark and Norway), Freedom of Information Act (USA). Similarly, in Austria, Section 3 (paragraph 5) of the *Bundesministeriengesetz* of 1973 imposes a statutory "duty to inform" on all Federal Ministries.

information more freely available to the public through legislative measures or administrative decisions.

The fundamental change from traditional practice is that a person requesting access to government information no longer has to show a "special interest". The right of a government agency to refuse access is restricted to a limited number of exempt categories of information specified in the legislation. However, in no country is there a total right of access to government documents.

For example, under the United States Freedom of Information Act, provision is made for exemption of documents the disclosure of which would prejudice national security, defence and international relations.[6] In other countries, such as Norway, exemptions are extended to include documents the disclosure of which would be reasonably likely to have a substantially adverse effect on national economic interests or expose commercial or financial enterprises unreasonably to disadvantage.[7]

In all of these countries, the approach taken to freedom of information legislation represents, and attempts to balance, legitimate and competing public interests. Efforts to secure political consensus on how best to achieve this balance have been difficult and time-consuming.

A number of arguments have been raised against such legislative proposals. For instance, fears that information disclosure would constrain administrative initiative and inhibit open, internal discussion of controversial issues were central to arguments raised by opponents in the United States to such legislation. It took almost 20 years of debate and compromise to overcome opposition before the Freedom of Information Act was passed in 1966 and another 11 years before the amended "Sunshine Act" came into effect in early 1977.

In those countries with Cabinet systems of government and long traditions of ministerial responsibility, opposition to such legislative reforms has been more entrenched. The fact that under such forms of parliamentary democracy, the Executive Government and Ministers are held directly accountable to Parliament, has meant that there has been considerable resistance to proposals that would threaten in any way the confidentiality of Cabinet deliberations and discussions. Moreover, traditional faith in government service neutrality and discipline has tended to reinforce beliefs that civil servants cannot offer advice freely and frankly if their views "can be distorted by the prism of partisanship into controversy and political attack".[8] Other fears have also been raised that the granting of such mandatory access to government information would prove to be costly, time-consuming, and possibly interfere with administrative efficiency.

The particular schemes adopted or proposed to promote such information disclosure reflect therefore, to a certain degree, different systems of government and political traditions in each country. The constitutional separation of powers and administrative structure that characterise the American political system and process also distinguish it from, for example, the Cabinet government and system of ministerial responsibility associated with Australia, Canada, the United Kingdom and the Scandinavian countries. These differences are reflected not only in the scope of legislation affecting government information disclosure, but in the

---

6. *US Freedom of Information Act,* Para. 552 (*b*)(1).
7. Norwegian *Law on Publicity in Administration,* Section 6(1).
8. Secretary of State Roberts, J., *Legislation on Public Access to Government Documents, Government Green Paper,* Secretary of State, Ottawa, June 1977. Also see: Attorney General's Department, *Policy Proposals for Freedom of Information Legislation,* Report of Interdepartmental Committee, AGPS, Canberra, November 1976; and *Hansard,* Australian Senate, 9 June 1978, for Second Reading speeches concerning the proposed 1978 Australian Freedom of Information Bill.

types of information exempted and in the administrative and adjudicatory procedures employed for its implementation.[9]

In countries where such legislation exists, more government information than ever before has become available to the general public. However, experience with such legislation has demonstrated that—contrary to some prior fears—public requests for information have not resulted in overburdensome administrative workloads. In fact, recent findings indicate a highly differentiated demand for such information. In the United States, most requests have come from government employees seeking access to personnel records, from businesses seeking information about competitors, and from lawyers involved in liability suits.[10] In Sweden and many other Scandinavian countries, the primary users of this right to access have been journalists, rather than the general public or interest groups.[11]

It would seem, therefore, that such mandatory information disclosure measures are, moreover, largely passive. A number of efforts have therefore been made to develop more active approaches. In Norway and Denmark, for example, National Information Services were established in 1965 and 1975 respectively, to co-ordinate the public information activities of government ministries. Although individual ministries in each country remain responsible for their own information matters, they consult this Service with respect to the selection of population target groups, use of information channels, and elaboration of advertisements. In addition, the Service also arranges its own advertisements, and the preparation of specialised government information features for presentation by the media.[12]

In a few countries, individual government departments and agencies also have their own information specialists and programmes. Many of these measures are exceptional and do not obtain in most OECD countries. For the most part, information disclosure remains largely a discretionary matter, especially with respect to administrative and regulatory tribunals. Even in those countries which have enacted freedom of information legislation, there are often many exemptions which provide government ministries and departments with ways to avoid compliance. Essentially political issues framed in terms of their narrow national economic interest or policy relevance can thus sometimes be exempted from mandatory access procedures.

One witnesses, therefore, in a number of countries increasing recourse to the courts as an effort to circumscribe the exercise of excessive administrative discretion and enforce existing disclosure laws. In one such celebrated case in the United States, the courts overruled an administrative decision to withhold a study prepared by the President's Science Advisor that was critical of the proposed supersonic transport. In presenting his decision that the SST report

---

9. For example, in Sweden and Norway minutes from Cabinet meetings are exempted from mandatory disclosure. In Denmark this exemption is extended to also include reports of meetings between Ministers as well as documents prepared specifically for such meetings. No such exemptions are provided for under the US Freedom of Information Act. Moreover, with respect to appeals, these are made directly to the courts under the US Act, whereas under proposed Australian legislation, now under consideration by the Senate Standing Committee on Constitutional and Legal Matters, they are referred to an Administrative Appeals Tribunal.

10. Kolata, G., "Freedom of Information Act: Problems at the FDA", *Science*, 189, 4 July, 1975, pp. 32-33.

11. Anderson, S., "Public Access to Government Files in Sweden", *American Journal of Comparative Law*, XXL, No. 3, Summer 1973.

12. Similar government information services are to be found in Germany, Austria and a number of other Western European countries.

be released, Judge Bazelon underscored the importance of informing the public in fields of science and technology:

> "The public's need for information is especially great in the fields of science and technology for the growth of specialised scientific knowledge threatens to outstrip our collective ability to control its effects on our lives... It would defeat... the purposes of the Act to withhold from the public factual information on a federal scientific program whose future is at the center of public debate."[13]

Even when such government information is accessible both in principle and in fact, the various interest groups do not have the same possibilities to avail themselves of such legislation. Groups must take the initiative to inform themselves, determine whether their own special interests are affected, and identify what further information is available and required. Some groups such as those representing private industrial interests have the financial and technical capabilities to do so. Others such as many citizen groups do not. The problem posed by such inequalities raises a number of questions concerning how to best promote equal access to information, whether or not government financial assistance ought to be provided to redress some inegalities, and how to ensure the more effective application of information disclosure laws and policies. These are issues that we shall address in detail below.

B. INFORMATION ON DECISION-MAKING

Information without knowledge about how decisions are arrived at—where a decision rests and when it will be taken—is often of limited value. The growth of government bureaucracy, the increasing complexity and intricacy of decisions, and the often unclear separation of decision-making responsibilities means that it is not always easy to discern the most appropriate or effective moment or means for involvement. Under such circumstances, information on the nature, scope and timing of decision-making becomes crucial to participation.

There is a long history of government efforts to provide citizens with the kind of information that will allow them to decide how and when to exercise their participatory rights. Some of these measures date from the last century —for instance the Norwegian Watercourse Act of 1897. This Act and the subsequent Administration Act stipulate that affected parties must be notified and consulted during the preparation of government regulations. Under the Norwegian system of "concessional treatment" (granting of permits), the Norwegian Watercourse and Electricity Board (NVE) must publish statements of intent by developers seeking project licensing approval, inviting participation from affected interest groups.[14]

Similar procedures are to be found in most countries. The *Journal Officiel* in France, the US *Federal Register,* and other such official organs serve as one means of informing the public on existing or new rules, government regulations and procedures concerning government decision-making. However, keeping up-to-date on notices appearing in, for example, the *Federal Register,* is often tedious and time-consuming since it contains more than 60,000 three-columned pages per year.

---

13. Wade, N., "Freedom of Information Officials Thwart Public Rights to Know", *Science,* 175, 4th February, 1972, pp. 498-502.

14. Garnasjordet, P.A. and Haagensen, K., *Public Involvement in Hydro-Electric Power Plant Planning,* Central Committee for Norwegian Research, Oslo, 1977, pp. I-VI.

Such journals also serve another function: that of eliciting public comment and measuring public attitudes. For example, the US Department of Health, Education and Welfare (DHEW) has in recent years published in the *Federal Register* early drafts of proposed guidelines to regulate the use of human subjects in research in order to solicit comments about its impending decisions, and to assess the degree of public acceptability of its procedure.

Over the last decade, other measures have been devised to inform the public about the details of impending decisions. The area of physical planning provides one example. Since the late 1960s, public concern in a number of densely-populated Western European countries over such issues as airport siting, motorway construction and physical planning in general have led to more active attempts to inform the public. In 1969, the Skeffington Report on physical planning in Britain recommended that the public be kept continuously informed in the preparation of local development plans, although it stopped short of advocating direct public participation in planning decisions.[15] In Belgium, Sweden and the Netherlands similar active approaches to informing the public on physical planning matters and the development of "structure plans" have also been adopted. While in Denmark, which has perhaps one of the most comprehensive national and regional physical planning systems, efforts have been undertaken to inform the public by means of public meetings and debates.[16]

One of the most significant advances in public access to information about technical decision-making is the requirement for the preparation of Environmental Impact Statements (EIS). Originally embodied in the US National Environmental Policy Act of 1969, an EIS must include a detailed statement of plausible environmental damage, feasible alternatives and anticipated losses in resource productivity on "all proposals for legislation and other major Federal actions significantly affecting the quality of the human environment".[17] Adopted by a growing number of governments, including Canada, Australia and France, these environmental assessment and review procedures have, more than perhaps any other single measure, opened up new channels of information and possibilities for citizen participation in technological decisions.

These various approaches to informing the public about decision-making procedures and participatory opportunities have for the most part developed slowly and incrementally. They reflect a growing realisation not only of the need for, but in some cases the practical benefits to be derived from, expanding information on public participatory opportunities. This appears to be especially the case with respect to administrative agencies and regulatory tribunals, which increasingly find themselves confronted with complex issues involving many different and competing interests. Their efforts to better inform the public on participatory opportunities is based in part upon the apparent recognition of the multiplicity of different "public interests" and on the pragmatic need to give consideration to those interests heretofore excluded or unrepresented.

Nevertheless, as with the case of mandatory access to information, these approaches are by and large passive and limited. Individual citizens or interest groups must assume the burden of informing themselves. Many are ill-equipped, technically and financially, to participate actively or effectively in decision-making processes. These same processes, to which we refer in detail below, are also

---

15. Ministry of Housing and Local Government (now Department of Environment), *People and Planning*, HMSO, London, 1969.
16. See: Mouritsen, P.E., *Public Involvement in Denmark*, Institute of Political Science, Aarhus University and the Danish Research Administration, Copenhagen, 1977, pp. 54-71.
17. Council on Environmental Quality, "Preparation of Environmental Impact Statement Guidelines", *Federal Register*, 38, 147, 1st August, 1973, pp. 20550-56.

often geared to consider testimony, generally of a highly esoteric or technical nature, which surpass levels of general public understanding. One result is that even well-informed citizens are often intimidated by the complexity and technicality of lines of argument and by the formality of court-like proceedings. Another is that, in taking recourse to legal representation, they become one step further removed from direct participation.

Thus, decision-making bodies traditionally accustomed to receiving technically-competent, legally-reasoned briefs often find it difficult to cope with qualitatively different types of testimony. Intervenors expressing strong social, political or emotional points of view are therefore often termed "technically incompetent" or, simply, "misinformed".

The other side of this coin, of course, is the obvious fact that many technologically-related issues obviously do contain technical as well as social and political aspects. One cannot easily reduce all discussion and debate to questions of social or political choice, and governments have come to recognise the need to promote broader public understanding in areas of scientific and technological controversy.

C. PUBLIC UNDERSTANDING

Government attempts to improve levels of public understanding on scientific and technologically-related matters take a number of different forms. Some of these include "study circles", national information campaigns, special non-governmental public information committees, as well as programmes designed to promote citizen self-education. Some are highly centralised with government assuming primary responsibility for their design and implementation, while others are more decentralised relying largely upon the efforts of local groups and non-governmental intermediary institutions.

Government ministries and departments with major scientific or technologically-related missions have, of course, been involved in the past in a variety of information activities aimed at informing the public about government programmes and initiatives. For example, the United States National Aeronautics and Space Administration (NASA) had, from its very inception in the early 1960s, undertaken programmes aimed at informing laymen about the benefits of space exploration, manned space flight, and aerospace-derived technology utilisation. Ministries and agencies responsible for the development of civilian nuclear power activities have likewise had a long history of involvement in this area. And, advisory bodies such as the US National Science Board (NSB) have been instrumental in the establishment of science information programmes and publications aimed at increasing public awareness of the actual situation, problems, and potentials of scientific research and its applications.[18]

What distinguishes many of these traditional activities and especially those of mission-oriented government agencies, from the newer forms of information programmes is their relative narrowness of scope. However, since the late 1960s, one witnesses an increasing trend towards the design and development of public information programmes which address more broad and often conflictual

---

18. Similar approaches are to be found in most OECD Member countries.

issues at the interface between science, technology and society. This shift is especially evident in the type of initiatives and information activities carried out, for example, by the Science Council of Canada, the Netherlands' Ministry for Science Policy and the recently established Swedish Research Councils Co-ordinating Board.

The most important distinguishing characteristic of more recent government-inspired and initiated public information programmes and campaigns is the degree to which they are purposefully aimed at encouraging broad-scale public discussion and debate. Some of the most recent examples are to be found in the area of energy policy and planning, and specifically in relation to civilian nuclear energy programmes. The example of nuclear energy is especially insightful in that it reveals the types of difficulties often encountered in attempting to inform the public in areas of extreme scientific and technical complexity and controversy.

1. Study Circle Mechanisms

The Swedish government's decision in late 1973 to undertake a major project of public education and consultation, represents one of the first comprehensive national initiatives to promote broader public understanding in the area of civilian nuclear policy and programme planning. The approach chosen was an ad hoc and decentralised one, based upon the mechanism of "study circles".[19] This is a system of small study groups, managed by the adult education associations, which are linked to the political parties and major popular organisations (trade unions, temperance groups, and religious organisations), and financed primarily by the State. Study circles are hardly new to the Swedish political scene; they date from the last century as a vehicle for the development of political democracy in that country.

The decisions to undertake this national education project and to employ this study circle mechanism were, however, closely intertwined. On the one hand, by late 1973, nuclear energy had become a pivotal, cross-party political issue. Public opposition to nuclear energy in Sweden had reached such a level by late 1973 that political party hegemony alone appeared insufficient to ensure adequate Parliamentary support for the government's proposed nuclear programme expansion. On the other hand, the study circles appeared to offer a means—consonant with Swedish traditions of seeking "consensus through compromise"—of defusing what was becoming a potentially volatile national political matter. Thus, the two central objectives of the government in undertaking its energy public education programme—that of broadening the base of decision-making and of establishing a consensus on future energy policy— appeared to be served by recourse to the study circle mechanism.

The approach taken by the government was to invite major Swedish social and political institutions to organise energy study circles, in return for which they received governmental financing and technical assistance.[20] The government did not attempt to intervene directly in the preparation for or conduct of the

---

19. Nelkin, D., *Technological Decisions and Democracy*, Sage Publications Ltd., London, 1977, pp. 60-64; and Frigren, S., "Public Education for Energy Policy Decisions", paper presented to the *International Conference on Nuclear Power and its Fuel Cycle*, IAEA, Salzburg, 2nd-13th May, 1977.

20. Ten organisations were invited to participate and seven accepted: the LO (major union) ran 3,000 circles; the ABF (run by the Social Democrats) sponsored 4,500 circles; two Folks Schools (run by the Centre and Liberal parties) sponsored 2,000; the Conservatives sponsored 500; and adult education groups run by the temperance movements and the Church of Sweden organised several hundred others: Nelkin, D., *Technological Decisions and Democracy, op. cit.*, pp. 61-62.

study circles, but provided each organising institution with a compilation of available official documentation on energy-related matters as well as information from opposition and environmental groups. Each sponsoring institution was responsible for developing its own documentation and received governmental financial assistance for these activities. In addition, the government established an independent "reference group" of scientific and technical experts, to which study circles could refer. And, a government publicity campaign was organised, aimed at encouraging public participation in the study circles, under the theme: "Learn more, and you will have more influence. Join an energy study circle".[21]

This particular study circle experiment lasted for but one year, with most circles completing their education projects by the end of 1974. An estimated total of 80,000 persons participated in the nearly 10,000 energy study circles organised. The majority of participants came from the already well-educated, well-informed politically-active population groups, not from the groups who were perhaps most in need of knowledge.[22]

Although the subjects treated by the study circles covered abroad range of energy-related issues (energy demand, use, alternative energy sources, safety), they most heavily emphasized nuclear energy. By the summer of 1974, the energy debate had become increasingly intense and polarised, owing in part to this education programme but also to the release of a number of committee reports concerning future energy requirements as well as articles in the press critical of energy planning. Thus, in the fall of 1974, the government saw the need to "stimulate a more factual and objective debate" by organising four public hearings in co-operation with its recently-established Energy Council.[23] These hearings were held between November 1974 and February 1975, involving representatives of both nuclear energy proponent and opponent groups, on problems of nuclear power, short-term conservation and supply prospects, and long-range energy policy alternatives.

When the government's energy bill was put forward in March 1975, it reflected both a more cautious approach to nuclear programme expansion and, to a certain degree, the increasing public concern about nuclear safety.[24] At the same time, the reports of discussions before the study circles had been fed back to the sponsoring associations and parties and forwarded to the politicians. In the ensuing Parliamentary debate, the *Riksdag* decided to approve the government's modest plans for nuclear programme expansion, and agreed to undertake a detailed review of nuclear energy policy again in late 1978.

The Swedish experience with the use of study circles as a mechanism for promoting a more informed public debate on nuclear energy has had decidedly mixed results. On the one hand, the reports from the study groups suggest continued and sometimes increased uncertainty and confusion. Moreover, subsequent surveys on the direct effect of the study circles on public attitudes, showed only slight differences in opinion between participants and non-participants.

---

21. Total estimated cost of the education project was about $650 000. In addition, the government organised four special seminars for journalists beginning in early 1974, to provide a technical briefing on nuclear energy-related matters.

22. A larger and more diversified sector of the public was indirectly reached through discussions on the job and at the home with participants who had participated directly in the study circles.

23. The Energy Council was set up in December 1973, comprised of 50 people selected to reflect diverse interests. This Council was represented in all four public hearings.

24. In addition to the 11 reactors already approved, two were to be added by 1985 at sites already approved. Conservation measures were proposed to achieve zero energy growth by 1990, and about $90 million was earmarked for energy R&D with $40 million going to conservation measures. Nelkin, D., *Technological Decisions and Democracy, op. cit.*, p. 64.

On the other hand, inquiries into the impact on attitudes of the four public hearings held in late 1974 and early 1975 indicated some shift in terms of enhanced public sympathy for the government position.

Nevertheless, this public education programme did result in a growing public awareness of the social and technical complexity of energy-related issues. The energy study circles acted as a catalyst for opening up a much broader public debate and interest in issues concerning the problems of large-scale technologies, ever-increasing energy consumption and economic growth, alternative energy sources, as well as relations between developed and developing countries.

The Swedish debate also revealed a breakdown in the historical left-right consensus on energy policy. Whereas the Conservatives and old-line Social Democrats were still able to raise a sufficient majority to ensure passage of the government's 1975 energy bill, they were opposed by the Centre Party, supported by some younger Social Democrats, the Liberals and by the Communists outside the Stalinist factions. Moreover, public opinion polls at that time (1975), indicated that rank-and-file supporters of all political parties, except those of the Centre Party with its anti-nuclear stand, were more concerned with the nuclear energy question than were the party leadership.[25]

This breakdown of the left-right consensus did not, however, produce a left-right cleavage on nuclear energy; it produced a cleavage running through every political party.[26] Nevertheless, public opinion poll findings up to April 1979 indicate a declining resistance to nuclear energy (Table 1).

The Centre Party which came to power in September 1976, partly due to favourable public support for its strong anti-nuclear stand, initially adopted a more conciliatory approach towards its two coalition partners, the Liberals and Conservatives, both of which support nuclear power. Given the dynamics of such a situation, the evolution in public sentiments towards nuclear energy, and the then apparent desire of the coalition government to avoid abandoning political power over a single issue, the "problem" has come to be perceived as less one of promoting public education per se, as to one of achieving political compromise.

To all intents and purposes, the use of the study circle mechanism as a primary means for informing the public on nuclear energy matters was "de-

Table 1
"IN A REFERENDUM, WOULD YOU VOTE
FOR OR AGAINST NUCLEAR ENERGY?"
Per cent

	October 1976	May 1977	September 1977	March 1978	September 1978
For	27	32	35	39	41
Against	57	49	46	40	37
Don't Know	17	19	19	21	22

Source: Sifo AB, Vällingby, Sweden.

---

25. Zetterberg, H.L., "Notes on Environmental Awareness and Political Change in Sweden", paper presented at the *Conference on Environmental Awareness and Political Change,* Wissenschaftszentrum, Berlin, 9th-10th January, 1978, p. 10.

26. *Ibid.* "Answers to opinion questions in the 1976 Swedish election also show a zero-correlation between position on the left-right continuum and attitude towards nuclear energy. Such zero-correlations do usually occur only on issues that are of little concern to the public. But the nuclear issue concerned the public a great deal and was a major issue in the 1976 election campaign."

activated", although some circles are still pursuing study programmes on energy issues as part of their normal activities. However, the government has made no provisions to continue or institutionalise this mechanism in the area of energy policy. Its general philosophy has been that, in the long run, "it must be the responsibility of the democratic society's ordinary educational system to provide its citizens with all the knowledge and information they need in order to be involved in the decision-making processes".[27]

Nonetheless, ad hoc efforts to provide the public with further information concerning nuclear energy have continued, though not without controversy. For example, in early 1977, the "Centrala Driftledningen" (CDL), a non-profit organisation promoting co-operation between Swedish electricity companies and half-financed by the State Power Board, published a number of booklets on the principles and problems of nuclear energy for distribution to schools, study circles, trade unions, companies and other organisations. Environmentalist groups were quick to criticise CDL's "study packets", characterising them as propaganda for nuclear power paid for with government funds.[28] At the same time, anti-nuclear groups renewed their demands for government funding, so as to allow them to compete on an equal footing with CDL and set up their own energy information service and support alternative energy studies. Thus far, no action has been taken on these demands.

During the spring of 1977, the Parliament passed the so-called "Conditions Act" requiring power companies to give proof of an acceptable contract for spent fuel reprocessing and to demonstrate that highly radioactive wastes could be disposed of in a safe manner. Thus, to an increasing degree the nuclear debate in Sweden had come to centre upon defining "acceptable" and "safe" in terms which are themselves socially acceptable. The coalition government headed by the Centre Party also appointed in early 1977 a Commission on Energy to examine the overall picture on energy supply and demand and the future of Sweden's nuclear programme. A year later, in late March 1978, the findings of this government Commission were released. Although no consensus was reached on the precise number of nuclear reactors to be built (ranging from 9 to 13), a majority of the Commission members recommended a much expanded nuclear programme, while leaving open the decision on future Swedish reprocessing plants.[29]

Nevertheless, in the spring of 1978, the Swedish Parliament approved plans for an expanded energy R&D programme oriented largely toward renewable energy sources, with special emphasis being placed on wind and biomass energy and on energy conservation measures. This three-year programme (1978/79-1980/81), which represents a major new national R&D initiative in terms of its size and focus on a set of specific technological goals, is expected to cost $183 million. The decision to put into operation Sweden's seventh recently-completed nuclear power plant is still pending. Moreover, public concern over the recent nuclear plant accident at Three Mile Island in Harrisburg (USA) has caused the government to adopt a cautious approach toward future nuclear programme development. A national referendum on nuclear energy is presently scheduled for the spring of 1980.

---

27. Quoted by Frigren, S., "Public Education for Energy Policy Decisions", *op. cit.,* p. 5.
28. Barnaby, W., "Spending a Packet", *Nature,* Vol. 268, 18th August, 1977, pp. 58-59.
29. Barnaby, W., "Swedish Experts Recommend Nuclear Energy—Without Reprocessing", *Nature,* Vol. 272, 23rd March, 1978, pp. 302-303.

In this brief description of the Swedish energy debate, a particularly interesting feature is the nature of its evolution and the continuous interplay between empirical facts and social values. One of the motivations behind the original decision in 1973 to open up public debate via the study circle mechanism was said to be to expose energy policy and planning to the diverse ideological viewpoints of political and social interest groups. It was recognised that because such energy-related issues had in the past been considered only within government departments, and largely in terms of technical questions, there was a need to broaden the base of decision-making. The study circle experiment in 1974 did not result in reduced public concern or uncertainty, but led to increased polarisation. This was reflected in the 1975 parliamentary debate on the government's energy bill and, to some degree, in the national election results of 1976.

In 1976 and 1977 the public education project was abandoned in favour of a return to the narrower confines of technical expertise of a government-appointed Energy Commission, whose purpose was not primarily one of public education but of scientific and technical forecasting and analysis. This occurred at the very moment that Parliament passed new legislation mandating that spent fuel reprocessing and disposal be conducted in an "acceptable" and "safe" manner: two normative terms requiring, above all else, a social consensus for their definition. This dilemma posed by the difficulty of integrating social values with technical expertise has been succinctly summarised by one Swedish member of Parliament as a central question facing contemporary democracy: "how to introduce the values a society wants to realise into a supposedly value-free examination of all possible courses of action open to that society?"[30]

2. Public Information Campaigns

Another mechanism designed for the promotion of broader public understanding in the area of nuclear energy policy and planning is the public information campaign. Austria, Denmark and Germany have developed considerable experience with this form of mechanism in recent years.

a) *German Bürgerdialog*

Perhaps the most well-known of these is the German Bürgerdialog or "dialogue with the citizens" on nuclear energy, a campaign conducted by the Ministry of Science and Technology (BMFT) since 1975. Preparation for this campaign began in late 1974, at a time when the German nuclear protest movement was just discernible but had not yet fully materialised. The German authorities saw the need to state and explain their attitude towards nuclear energy in a democratic way. In the light of the 1973 oil embargo, they recognised the importance of promoting public information and discussion on future energy demand and its relation to economic growth and energy utilisation, as well as of achieving a broader consensus on the requirements, security and reasonable proportion of nuclear power as an energy source.[31]

---

30. Ms. Birgitta Hambraes, Centre Party Member, quoted in *Nature,* Vol. 272, 23rd March, 1978, p. 303.
31. See: Lang, K., "Information on Nuclear Energy in the Federal Republic of Germany: Establishment of a Dialogue Between the Public and Government Authorities", IAEA, CN 36/81, Salzburg, 1977.

This initiative represented the first attempt of the Federal government to discuss the introduction and development of a new, large-scale technology with the public. Consequently, its initial approach was a modest and cautious one. Beginning in 1975, the government initiated a major publicity campaign aimed at informing the public as to the views of government with respect to: the need for energy questions to be discussed with the general public; the need for nuclear energy and its safety; R&D requirements in the field of alternative non-nuclear energy sources; energy conservation measures; radioactive waste disposal; and the role of citizens' initiative groups.

Eight public advertisements were distributed along with return-coupons allowing citizens to order additional information material and receive notification of subsequent discussion meetings. The Ministry of Science and Technology (BMFT) also published three technical handbooks: "Nuclear Energy - Information for the Citizen" ("Kernenergie - eine Bürgerinformation"); "Discussions and Interviews on Nuclear Energy" ("Gespräche und Interviews zur Kernenergie"); and "Information Letter on Nuclear Energy" ("Informationsbrief Kernenergie"). Over a million copies of these books and other related material have been issued to date. The total cost of the activities during the first year was about $300,000.

Beginning in 1976, 17 public seminars and discussions were organised in which the then Minister for Science and Technology, Hans Matthöfer, himself often participated, along with representatives from the electric power utilities, public interest groups, and the general public.[32] Nearly 4,000 people took part in these discussions which, while focused primarily on questions concerning nuclear energy, also began at this time to address more general issues of German energy policy. In addition, efforts were made to encourage various social organisations to broaden their involvement by sponsoring their own discussion meetings and study groups.

These efforts were intensified in 1977, and steps were taken to decentralise campaign activities and expand the base of discussion. Adult education centres, political parties, trade unions and Church groups were encouraged to organise "opinion-forming processes" and discussions within their respective organisations. BMFT agreed to provide technical and financial assistance to these groups. At the same time, it organised a further 20 seminars aimed at specific "target population groups" and prepared an additional set of energy "information packages" for dissemination to schools and adult education centres. A total of about $1.4 million was allocated each year to these information activities in 1976 and 1977.

During 1978, further efforts were made to stimulate fuller discussion especially by the political parties and the large employer and trade union organisations such as the BDI (Association of German Industrialists) and the DGB (the Trade Union Congress). Expenditures for 1978 were estimated at more than double those for each of the two previous years—about $3.2 million, with one-third allocated to government information activities and more than half to individual groups and associations. The aim of this "dialogue" was to expand the scope of public discussion on more general energy-related matters and to further decentralise "opinion-forming" processes.

When this public information campaign began in early 1975, it was quickly denounced by nuclear opponents as a government "pro-nuclear propaganda campaign".[33] However, as it became evident that the government had the

---

32. See: BMTF, *Kernenergie — eine Bürgerinformation*, BMTF, Bonn, 1976.
33. Lang. K., *op. cit.*, pp. 132-133.

intention of including the opinions of nuclear energy critics in the discussions and of providing opportunities for opponents to present their views and counter-arguments, criticisms of the "dialogue as monologue" began to fade. Industrial proponents of nuclear energy also were initially sceptical of this "dialogue", for a number of opposite reasons, but subsequently agreed to participate in discussions and present their views.

The "dialogue with the citizens" did little, however, to impede or contain the growth of a strong nuclear protest movement. In November 1976 and March 1977, major public demonstrations took place against plans for the siting of nuclear power plants at Brokdorf in Schleswig-Holstein and Grohnde in Lower Saxony. These protest movements focused primarily on environmental-impact concerns and the adequacy of reactor containment measures. In both instances, construction permits previously approved by the government were subsequently denied in separate administrative court decisions.[34]

The information campaign does not appear to have influenced heavily public attitudes toward nuclear energy one way or the other. These have remained relatively stable over the last three years, with approximately 15 to 20 per cent of the population fairly strongly "for" or "against" continued dependence on nuclear power as an energy source, and a significantly large 60 per cent more or less undecided.[35] Nevertheless, the campaign has apparently served to channel discussion towards some of the key problems confronting German energy policy-makers and the body politic as a whole, namely the issues of nuclear waste disposal and fuel reprocessing. It has also opened up a more general debate on the relationship between future energy policies and economic growth.[36] Government estimates of future energy demand (by 1985) have, as in most OECD Member countries, been drastically reduced in recent years, which has permitted an added degree of flexibility and time for reconsidering plans for the expansion of nuclear-electric generation facilities.[37]

The Social Democratic Party (SPD) at its annual Congress in November 1977, achieved a compromise position—i.e., to maintain the operation of existing nuclear plants; to complete construction on plants already approved; but to delay further expansion until the problems of spent fuel reprocessing and waste disposal are resolved. The SPD's coalition partners, the Liberal Party, adopted a similar compromise position.

In Germany, as in most other countries with major civilian nuclear-electric energy programmes, spent fuel reprocessing and highly radioactive waste disposal have therefore become perhaps the key technical and political issues affecting the pace of future nuclear programme development. After a number of detailed studies, the Federal government proposed that a spent fuel reprocessing plant and storage facility be built in Lower Saxony; a state in which available salt

---

34. Similar administrative court decisions also resulted in suspension of construction activities at Whyl on the Lower Rhine in Baden-Württemberg and at the Mülheim-Kärlich plant in the Rhineland-Palatinate. See below pp. 92-93.
35. Lang, K. and Popp, M., *op. cit.*, p. 8.
36. The title of the information campaign undertaken in 1977 highlights this change: BMFT, *Kernenergie und ihre Alternativen* (Nuclear Energy and its Alternatives), BMFT, Bonn, 1977.
37. The current estimated energy demand (1977-1978) is 370 million tons of coal equivalent (Mtece). It was originally estimated in 1973 that energy demand in 1985 would amount to 610 Mtece. This was reduced to 496 Mtece in early 1977 and again to 480 Mtece in November 1977. An independent estimate made by the Institute for Applied Systems Research and Prognosis (ISP) has revised these energy demand figures downward to 437 Mtece. *Nuclear Engineering International,* January 1978, p. 23; and *Nature,* Vol. 272, 30th March, 1978, p. 393.

dome formations were considered to provide the most suitable geological sites in Germany for "final storage" of highly radioactive wastes. However, given the considerable autonomy of individual states in the Federal Republic, Länder approval is necessary for the implementation of these Federal plans.

The state government of Lower Saxony, under its Premier Ernst Albrecht, initially rejected the three alternative sites proposed by the Federal Government, suggesting instead that the facility might be built at Gorleben. However, Premier Albrecht, in part in response to opposition from some local residents and environmental groups, established in June 1978 a panel of international "critics" to review Federal government plans for the facility.[38] This so-called Gorleben International Review Panel recommended in the spring of 1979 that the spent fuel reprocessing and storage facilities not be built. The decision on whether or not to go ahead with construction has been postponed.

It is interesting to note that the Austrian and Danish approaches to the design and implementation of nuclear energy public information campaigns represent two different versions of the German Bürgerdialog model. Compared to the German campaign, the Austrian approach was more structured, and centralised under the direction of the Federal Ministry of Industry. The Danish campaign, on the other hand, was highly decentralised and administered by an independent non-governmental commission which emphasised local self-education. Unlike the on-going German "dialogue with citizens", both the Austrian and Danish nuclear information campaigns were terminated after a relatively short period of operation.

b) *Austrian Energy Information Campaign*

The Austrian decision in 1968 to construct its first nuclear power plant at Zwentendorf met with little local or national resistance. It was not until plans for a second plant in Sankt Panthaleon were announced in 1973 that opposition began to mobilise. Between 1973 and 1975, a number of anti-nuclear demonstrations took place in Austria, some focusing on the proposed Sankt Panthaleon site, others on plans by the Swiss government to locate a nuclear power plant several kilometres from Austria's western border. These incidents received considerable media coverage. And when, in April 1975, one of Austria's more influential newspapers organised a nationally-televised debate between nuclear energy proponents and opponents, the Austrian anti-nuclear movement achieved a certain national notoriety. Several months later, Chancellor Bruno Kreisky, who had himself participated in these public discussions, announced that such a controversial and important issue as nuclear energy required broader public discussion and could not be treated by experts alone.[39]

---

38. The state government's decision was unique in that it was the first time a European government had invited a panel of foreign experts to assess a national civil nuclear programme. This review panel includes ten experts from the United States, four from the United Kingdom, four from Sweden and two from France.

39. Nelkin, D., *Technological Decisions and Democracy, op. cit.,* pp. 53-56.

During the first half of 1976 the Ministry of Industry began preparations for a nationwide public information campaign. One of the central objectives of this campaign was to try to achieve a broader political as well as public consensus on government plans for the construction of three nuclear plants by 1990.

The first step taken was to invite technical experts from both proponent and opposition groups to provide lists of the key questions and concerns that they felt must be considered prior to arriving at firm decisions on the nuclear programme. These questions were then structured in terms of ten principal themes, and experts were assigned to develop available information on each theme. At the same time, a glossary of technical nuclear energy terms was assembled and published as a dictionary for distribution to all Austrian citizens.

Nationally-televised public debates were then organised focusing on issues and questions upon which a sufficient technical and scientific consensus did not appear to exist. Proponent and opposition experts were invited to lead these discussions and to respond to written questions submitted by the public. An effort was made to ensure an unbiased representation of technical expertise by excluding from these debates experts employed by industrial firms involved in the Austrian nuclear programme. These debates were broadcast from October 1976 until the late spring of 1977.

From the beginning, the nuclear opposition groups which had in general supported the idea of the campaign during its preparatory phase, intervened in these televised debates in a perhaps unexpectedly overt and disruptive manner. At the first meeting in Linz, where opposition group members out-numbered nuclear proponents, the audience demanded that the Chairman be replaced and the debate reoriented to issues other than those proposed for discussion. Similar "disturbances" were also experienced in subsequent debates, as opposition groups sought to further exploit for their own purposes the opportunity provided by the televised information campaign.

It is difficult to fully assess the educational value of this campaign or its impact on public attitudes towards nuclear energy.[40] As in the case of Sweden and Germany, it clearly served to increase public awareness and open up a broader debate on overall Austrian energy policy. However, it also served to greatly politicise the issue. Although opinion poll findings from before and just after the information campaign indicated a significant growth in public opposition to nuclear energy, by late 1977 popular opinion appeared to be evenly split on the issue. By early 1978, however, opinion poll findings indicated a slight majority of the population in favour of the government's nuclear programme.

Chancellor Kreisky announced shortly thereafter that a national referendum would be held. On November 5, 1978, the Austrians voted against government plans to put into operation Austria's first nuclear power plant at Zwentendorf. Although the referendum was to have had an advisory and not a binding effect on Parliament's decision, the Austrian Parliament subsequently voted not to commission the nuclear plant, thus effectively terminating Austria's nuclear programme.

---

40. For a preliminary assessment of the impact of the campaign on public attitudes, see: Hirsch, H. and Nowotny, H., "Information and Opposition in Austrian Nuclear Energy Policy", *Minerva*, Vol. XV, Autumn-Winter 1977, pp. 314-334. Helga Nowotny, *Kernenergie: Gefahr oder Notwendigkeit*, Surkkamp, Frankfurt, 1979.

c) *Danish Energy Information Campaign*

The evolution of the Danish nuclear energy debate does not differ greatly from that of many other western industrialised countries, except perhaps in the speed with which nuclear energy has become a major national political issue. Although atomic energy research has been underway in Denmark since 1955, it was not until late 1973 that plans for the construction of Denmark's first nuclear power plant were announced.[41] This coincided with a moment of intense public preoccupation with energy matters, following the 1973 "oil crisis". Within a span of six months, the nuclear issue became a topic of considerable national attention and debate.

In April 1974, the electrical utilities announced the selection of Gyllingnaes, near the city of Aarhus, as the site for their first reactor, and began preparations for a major public information campaign aimed at informing the local citizenry as to its reasons for selecting this site. During the same period, a citizen activist group calling itself the Organisation for Information on Nuclear Power (OOA) which had been founded three months earlier in response to utility plans for an accelerated nuclear programme, began organising regional and local groups.[42] During the same month, the Danish Parliament (Folketinget) held its first major energy debate and decided to establish a special committee on energy policy. In the course of this parliamentary debate, several parties called for broader public discussion on energy questions, and spokesmen for the ruling Social Democratic party suggested the need for a major public information campaign.

Therefore, in June 1974, the Minister of Industry and Commerce established a special non-governmental Committee on Energy Information (Energi Oplysning Udvalget - EOU) for this purpose.[43] This Committee consisted of representatives from the general education organisations, libraries and folk high schools, but not from the Ministry itself. The EOU was given considerable leeway to determine its own course of action, strategies and materials.[44] The only direct control exercised by the government was via its Finance Committee, which appropriated approximately $410,000 for the two-year campaign.[45]

This was not the first time that the Danish government had undertaken a major public information campaign effort. Just two years earlier, it had allocated more than $2 million for the organisation of an information campaign in connection with the national referendum on Danish membership in the European Economic Community (EEC). During this campaign the government provided block grants primarily to the major Danish educational associations, but also to political parties and other social organisations for the conduct of courses, study circles, lectures and seminars. The experience with this particular campaign

---

41. Work had been carried out at the Danish atomic energy research plant, Risø, in the middle 1960s on a heavy water reactor prototype which was subsequently abandoned in 1970. Also, an industrial consortium, the Dansk Atomreaktor Konsortium, had been formed for the production of reactors, but these plans were never realised.

42. The OOA (Organisationen til oplysning om Atomkraft) had organised 15 local groups by March 1974, 30 by August 1974, 90 by early 1976, and 160 by June 1977. Mouritsen, P.E., *Public Involvement in Denmark, op. cit.*, p. 103.

43. It was felt that a purely government-organised campaign or one conducted under the auspices of Risø, the atomic energy research centre, might be criticised as being unduly biased in favour of nuclear energy; thus, the decision to establish a non-governmental commission.

44. The government mandate only stipulated that the EOU should inform the public about overall energy problems and especially "about specific conditions that in several respects are attached to possible construction of nuclear power plants in Denmark...", *Lov om energipolitiske foranstaltninger*, No. 194, 28th April, 1976.

45. The actual campaign did not begin, however, until November 1974 and lasted for 1½ years only.

was viewed by many people, both in and out of government, as having been positive—at least from the point of view that the pro-membership faction won the EEC referendum by a landslide.[46]

The Energy Information Committee (EOU) did not, however, choose to follow this "block grant" approach. Neither did it consider the Swedish study circle model appropriate nor the type of information material used in that campaign sufficiently representative of different points of view. Rather, it sought to implement an approach responsive to the following three principles:

 a) that the energy issues be handled and examined primarily in terms of their political and social context, not solely in technical terms;
 b) that parity be sought between "pro" and "con" positions in the presentation of all material produced by the Committee; and
 c) that the campaign should seek to improve the possibilities of public participation in the political decision-making process, either directly or indirectly by influencing elected representatives at different levels.

This latter point implied an emphasis upon encouraging the growth of locally-based information activities.

The Danish approach to informing the public therefore focused heavily upon encouraging "grass roots" initiatives through the distribution of grants to a large number of individual regional and local groups. Of a total of $165,000 allocated for these activities, only a relatively small amount was distributed in the form of block grants to educational associations. Most of the funds went to underwrite the cost of local meetings, study circles, and other self-education activities. The Committee did not demand rigorous accountability as to how these funds were used as long as the subsidised activities met EOU's basic conditions for comprehensiveness and public accessibility. In some instances, local initiatives focused primarily on studying alternative energy sources and not nuclear energy matters. An estimated 150,000 people participated in this phase of the campaign.

In addition, the EOU itself produced, with assistance from outside experts, a considerable amount of technical and policy-related information material for distribution to the general public. This included the publication of six "basic books" on general energy-related matters. However, unlike the German or Austrian campaigns, nuclear energy per se did not figure predominantly in these books.[47] The Committee also sought to establish close contacts and working relationships with media representatives, although the approach taken does not appear to have been as structured as the Swedish one.

It is obviously difficult to establish a direct correlation between increased public awareness generated by this campaign and changes in public attitudes towards nuclear energy in Denmark. However, one set of public opinion poll findings (Table 2) on public voting preferences in the event of a national referendum indicates a steady growth in opposition to nuclear power since completion of the public information campaign.[48]

---

46. For detailed discussion of this government-financed campaign, see *Oplysningsarbejdet om EF 1972*, Undervisningsministeriet, Sekretariatet for oplysningsvirksomhed om Faellesmarkedsproblemer, 1973.
47. These were on: Danish energy policy, nuclear power, energy economy and planning, energy and the environment, economic and political aspects of natural resources, and renewable energy sources.
48. Similar opinion polls have been undertaken by other institutes, with some deviations noted with respect to absolute figures. However, the general trends indicated in all polls are consistent.

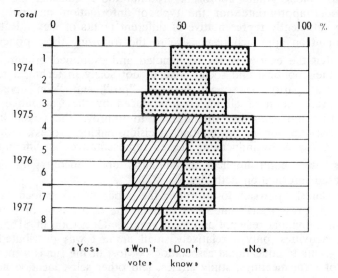

Table 2
QUESTION: «IF THERE WERE A REFERENDUM TODAY, WHAT WOULD YOU VOTE?»

Source: Observa Institute and the *Jyllands-Posten*, Copenhagen.

The Danish public information campaign did not, however, take place without strong criticism being raised by some members of the nuclear research community, the pro-nuclear Federation of Danish Industries, or the more conservative political parties. EOU's six "basic books" came under especially severe attack for their alleged lack of objectivity. Perhaps the major criticism voiced about EOU's handling of the energy campaign was that the Committee had too advertently intertwined technical with political and social aspects. This criticism was raised in Parliament by one leading member of the Danish Liberal Party in the following terms:

"We have been taught how a technical and scientific approach can be politicised... I remind you that this part of the debate also originates... from people who... think that technical-scientific questions are nothing concrete... but something to be decided upon politically, to be the subject of a debate, something subjective. If pushed to extremes, we will be where we have been, on the outskirts of the debate which has been going on... and what I regret most (is that) owing to misleading information (many people), have encountered great difficulties and had a lot of trouble; this applies really both to people inside and outside this House."[49]

It is not easy to discern the degree to which this public information campaign did, as claimed, engender greater public confusion or uncertainty. Public opinion poll findings would appear to contradict this allegation. During the period 1975-1977, the number of people polled who were undecided or who said they would abstain from voting in a nuclear referendum remained relatively constant at

---

49. *Folketingstidende*, 1975-1976, Forhandlinger, sp. 10124. Quoted in Mouritsen, P.E., *Public Involvement in Denmark*, op. cit., p. 114.

approximately 30 per cent (Table 2). Moreover, as suggested by subsequent analyses, some of those most critical of the EOU campaign for having "intertwined" social and political with technical aspects were the very same people who favoured nuclear programme development and who sought to contain the debate to the narrower confines of purely scientific and technical matters.[50]

However, what the campaign did to a large extent achieve was an increase of awareness among the general public with respect to alternative energy strategies and to possible social and economic consequences of expanded nuclear programme development. It also served, as in the case of Austria, to further politicise an already highly political debate.

When, in May of 1976, the government put forward its long-range energy plan, which envisioned the construction of five nuclear plants by 1995, the Parliament was increasingly divided on this issue.[51] This was also the case within several major political parties. As the Social Democratic Prime Minister Anker Jørgensen noted in August 1976:

> "We would rather not have the Social Democratic Party exposed to such a split-up for something so emotional. It might make it impossible for the Social Democratic Party to pursue day-to-day policy in the coming years. This is the reason why we are going to take it easy and not make hasty decisions without a strong backup in the population."[52]

Several weeks later, the government decided to postpone indefinitely parliamentary action on the Danish Execution Bill, an act, which if passed, would have paved the way for initiating revised authorisation procedures for the construction of Denmark's first nuclear power plant. The official reason given for this decision was that further investigations were required with respect to the problems of nuclear waste storage and public investment planning before a decision could be made.[53]

Nevertheless, despite the recognition of the fact that, in the Prime Minister's words, "many people feel that they have not been sufficiently informed and... are emotionally affected by the uncertainty in connection with the use of nuclear power",[54] the government information campaign was abolished. It was believed that more "traditional" channels of information would be adequate to meet the public's need for such information.

Seen in retrospect, there was certainly some truth in this assessment. For, one of the things that the government information campaign had achieved was not only increased public awareness, but more active public engagement in activities concerned with energy-related matters in Denmark. For example, membership in the activist OOA (Organisation for Information on Nuclear Power) has grown considerably: the number of local groups has grown continually since 1974, covering now most regions in the country. OOA used some of the grants it received from the EOU campaign to finance its own membership expansion,

---

50. *Ibid.*, pp. 114-115 and 140-144.
51. Of the 11 political parties represented in Parliament in 1977, five parties (Liberals, Conservatives, Christian People's Party, Centre Democrats, and Progress Party) were strongly in favour of the nuclear energy programme. They held altogether 79 of the 175 seats. The Social Democrats (65 seats) were split on this issue as was the Radical-Liberal Party (6 seats). The Socialist People's Party, Single Tax Party and Left-Wing Socialists (18 seats combined) were in firm opposition on the issue. The Communist Party (7 seats) is in favour of the nuclear energy programme in principle, but opposed to the programme in fact, owing to what it judged to be insufficient public controls over facility operations.
52. Quoted in *Metal*, No. 9, 1976, pp. 20-21.
53. The so-called "Execution Bill" was a form of enabling legislation whose passage was required before the Danish Nuclear Installations Act, previously approved by Parliament, could come into effect.
54. Quoted in *Metal, op. cit.*, p. 20.

but also to develop its own expertise in many energy-related areas. It is considered by many today be one of the more effective Danish "grass roots" organisations engaged in the promotion of broader public understanding in the area of energy policy and alternative energy sources.[55]

The electrical utilities have also become heavily engaged since 1974 in trying to promote a more informed public debate, especially with respect to nuclear energy. The major association of electricity utilities (ELSAM), partly in response to the government campaign, has increased its own information campaign, providing considerable publicity for radio and television and detailed energy papers and booklets to the general public. It has also focused heavily upon better informing journalists and teachers by means of special lectures and seminars conducted by its own technical personnel.[56]

In addition to these groups, a new organisation called Reliable Energy Information (Reel Energi Oplysning - REO) was founded in late 1976, with the aim of "creating broad popular understanding of the desirability of introducing nuclear power in Denmark by procuring objective and correct information on energy and on nuclear energy power in particular".[57]

To a large extent, then, the government-initiated public information campaign has been responsible for the growth in the number of local and industrial-based organisations, which have in turn themselves assumed responsibilities for providing the public with information on energy-related matters. These latter information activities are, of course, not solely aimed at promoting a more informed public debate, but also at engendering support for each organisation's own particular goals and policies. Thus, while the Danish information campaign may not have achieved a social or political consensus on future national energy policy, it has nonetheless contributed in an important way to the development and diversification of alternative sources of information upon which to base that policy.

These examples of how Sweden, Germany, Austria and Denmark have sought to promote broader public understanding in the area of energy policy and nuclear programmes are, of course, not the only experiences that could be cited. France and Canada provide two other rather different examples.

d) *French Nuclear Information Council*

France differs from most other OECD countries in that it is the only one without any overall atomic energy legislation. All major decisions with respect to nuclear policy and programme development have traditionally been taken by the national administrative authorities. The Parliament, up to the present day, has never voted specifically on the question of nuclear power. During the 1950s and 1960s, when Parliament adopted the first series of successive five-year economic plans, it tended to consider nuclear energy programme strategies as

---

55. See: Mouritsen, P.E., *op. cit.*, pp. 122-132.

56. ELSAM, not unlike electrical utilities in many other countries, has had some difficulty in adapting to exigencies imposed by public demands for more direct consultation and participation in energy decision-making. For example, as one ELSAM representative commented, "... in my opinion, it ought to be those directly dealing with things that make decisions on the introduction (of nuclear power plants)... that means the ELSAM and the KRAFTIMPORT. In fact, our respective managing boards". Quoted in *Jyllands-Posten*, 6th June, 1976.

57. REO, which had about 450 members by mid-1977, is generally considered to be a counterpart to the OOA.

but one set of activities among many; all essentially aimed at the same general objective of promoting national, social and economic development and modernisation. Nuclear energy was not singled out for special treatment.

As in most industrialised countries, public concerns over plans for the development in France of civil nuclear energy did not begin to be voiced until the early 1970s. The first major public demonstrations against nuclear plant siting took place during the summer of 1971 at Bugey and Fessenheim. However, only after the spring of 1974, when the government announced its intentions (under the Messmer Plan) to rapidly expand its nuclear reactor construction programme to 40 units (40,000 MWe) on about 15 sites by 1985, did this issue begin to receive wide public attention. When, in April 1975, the Parliament began its first detailed debate on this programme, one of the key issues of controversy concerned the fact that the major political decisions had already been taken and ratified without recourse to broad public debate.[58]

Since then, public opposition to nuclear power has grown considerably. Nevertheless, significant public opinion poll findings indicate that a majority of the French public still support the government's plans for nuclear programme expansion.[59] Nuclear opponents have tended to concentrate in recent years mainly on issues involving the development of fast breeder reactors, a controversy especially marked by massive demonstrations at Creys-Malville in the summer of 1977. Opposition is also very much in evidence with respect to local siting decisions for conventional nuclear plants, but owing to the rather particular nature of the French decision-making tradition, many regional and local opposition groups have had relatively few means at their disposal, short of overt opposition, for directly influencing decision-making.

Given the highly centralised structure of French public administration, most policies and decisions are adopted and implemented at the national level. Regional and local consultation procedures do however exist, and over the last two years the national government has adopted a number of new measures aimed at broadening these consultative approaches, while at the same time retaining its final decision-making prerogatives. These include provisions for reforms of existing public interest inquiry procedures and the requirement that environmental impact studies be carried out prior to the undertaking of major development projects.[60]

The issue of public access to information concerning risks associated with nuclear energy has also received increasing attention. Nuclear opposition and environmentalist groups, and most notably the French section of Friends of the Earth and the Scientists' Group for Nuclear Information (GSIEN), have argued that the public has not been fully or adequately informed about the actual or potential risks associated with nuclear programme expansion and that the government's own approach to risk assessment has been inadequate.[61] These groups

---

58. The debate had no direct bearing on policy in that it was not followed by a vote. However, this debate indirectly served the interests of the nuclear protest movement in that it received considerable media coverage, and thus awakened some of the public to energy policy problems and industrial hazards.
59. See: Dumenil, G., "Energie nucléaire et opinion publique", Report No. 2, CORDES-CNRS Contract on "Le Débat nucléaire en France". See also proceedings of the symposium on "Les implications psychosociologiques du développement de l'énergie nucléaire", Paris, January, 1977. Francis Fagnani, *et al., Nucleopolis,* Editions de l'Université de Grenoble, 1979.
60. *Journal Officiel,* 13th October, 1977.
61. See: Groupement de scientifiques pour l'information sur l'énergie nucléaire, *Electronucléaire: Danger,* Le Seuil, Paris, 1977.

have sought by various means to draw public attention to these concerns.⁶² At the same time, the nationalised electrical utility, Electricité de France (EDF), has mounted its own public information campaign aimed principally at assuring citizens of the general acceptability of these risks and of the necessity for expanded nuclear programme development.

The government has sought to ensure a more open and "rational" discussion of these different claims and counter-claims. It has, however, hesitated to adopt the kind of national public information campaign mechanism of Germany, Denmark or Austria, fearing perhaps that such an approach might precipitate the very kind of emotional controversy it is seeking to avoid. Rather, the government has adopted a mechanism somewhat akin to the concept of a "science court" as proposed in the United States, for the more "objective" and "rational" review and discussion of opposing viewpoints as to the adequacy of existing technical information.

For this purpose, the government established in early 1978 a national Information Council on Nuclear Energy (Conseil de l'information sur l'énergie électronucléaire), comprised of 18 members representing diverse interests, and presided over by the Minister of Health and Family.⁶³ This Council has a broad mandate not only to see that "the public has access to information on questions relevant to nuclear energy", but to advise the government as to the conditions on which public access to information should be granted, and to propose forms and means for the diffusion of such information.⁶⁴ At the same time, however, the Council is not engaged in the actual provision of information to French citizens; such functions are presently performed by governmental ministries, the state-owned power utility and other related government institutions. It is still too early to judge the significance of the Council's role in promoting broader levels of public understanding.

e) *Canadian Committee on Nuclear Issues*

In Canada, which ranks fifth in the world in terms of nuclear-electric energy production, the nuclear debate has developed with a certain retard.⁶⁵ Its first commercial reactor went into operation in 1967 and since then up until the mid-1970s there was relatively little public interest, concern or opposition expressed with respect to Canada's nuclear programme development, which currently includes 22 reactors in operation, under construction or already committed.⁶⁶

---

62. In addition to their own group information activities, discussions and meetings, various environmentalist and counter-culture groups also publish magazines and newspapers, such as *Le Sauvage, Survivre et vivre, Charlie hebdo* and *Gueule ouverte*. In addition, a number of books have also appeared recently, including: Colson, J.P., *Le nucléaire sans le Français,* Maspéro, Paris, 1977; Lenoir, Y., *Technocratie française,* J.J. Pauvert, Paris, 1977; and Puiseux, L., *La Babel nucléaire,* Editions Galilée, Paris, 1977.
63. This Council includes: four elected officials; six representatives from major environmentalist groups (including the secretary-general of Friends of the Earth); four members from different French science and medical academies; and four "qualified persons from the fields of energy, economics and telecommunications, chosen by the Prime Minister". *Journal Officiel,* 11th November, 1977.
64. *Ibid.*
65. The four leading producers of nuclear-electric energy are the United States, United Kingdom, Germany and France.
66. This nuclear capacity is largely concentrated in the Province of Ontario, which has 19 reactors in operation, under construction or already committed. It has recently been found that seven of these reactors will have to be closed down, prematurely, for one year before 1985 in order to replace their heavy water cooling systems. *Le Monde,* Paris, 18th August, 1978.

Nonetheless, a major siting controversy over the construction of a 600 MWe nuclear plant at Point Lepreau in the Province of New Brunswick in 1974 received considerable national attention. In 1975, a number of relatively small anti-nuclear groups appeared and, joined by existing environmental organisations, formed an umbrella organisation calling itself the Canadian Coalition for Nuclear Responsibility (CCNR).[67] But it was not until 1976 that overt opposition and agitation really began to manifest themselves in Canada. These opposition groups have since become increasingly concerned with matters relating to nuclear fuel recycling and high-level radioactive waste disposal. CCNR has called for a national moratorium on nuclear programme expansion until these issues are more publicly aired and adequately resolved. Opposition groups have also increased their demands for a nationwide inquiry into Canada's nuclear energy programme.

As in many other countries, Canadian levels of public understanding on nuclear matters are generally low. One national survey undertaken in 1976, showed that nearly half of the population (44 per cent) were not even aware that nuclear power could be used to generate electricity.[68] Apart from expressing a certain fear of nuclear technology, many Canadians also assign a low credibility to the institutions responsible for nuclear power.[69] Institutional sources of information concerning nuclear-related issues are, according to these survey findings, viewed as generally less credible than those emanating from scientific and technical experts.[70]

In early 1978, the Federal government established a Committee on Nuclear Issues in the Community in response to the need to promote broader public understanding of nuclear-related issues, but also to serve the specific information needs of residents in those communities where nuclear power facilities are expected to be located. This Committee was organised under the auspices of the Royal Society of Canada and the non-governmental Science Council of Canada, which provides secretariat assistance. It had an initial operating budget of $200,000.

The activities of this Committee got underway in the late spring of 1978, but not without some difficulties. During its first community meeting held at Thunder Bay in Ontario, disturbances were caused over the fact that the Committee had initially decided that the meeting should be "closed" to all non-residents of the Thunder Bay area. After considerable dispute and some misgivings, the Committee agreed to open the meeting to all interested parties. Shortly thereafter one member of the Committee resigned, claiming undue government manipulation of this independent non-governmental Committee. These episodes were widely reported by the media, and opposition and activist groups such as CCNR and Energy Probe charged that the Committee would turn out to be simply a propaganda agency for nuclear power. These groups reiterated their demands to the Federal government for "equal funding" so as to be able to present, in their words, "the other side of the debate". The Committee subsequently agreed to expand its membership to include two representatives from these opposition groups and one representative from the media, in an attempt to both bolster its own credibility and to ensure a more balanced representation of different viewpoints.

---

67. CCNR is comprised of more than 45 individual groups.
68. Greer-Wootten, B. and Mitson, L., *Nuclear Power and the Canadian Public*, Institute for Behavioural Research, York University, Toronto, June, 1976, p. xiii.
69. Davies, J.E.O., *et al.*, "Canadian Attitudes to Nuclear Power", paper presented to the *International Conference on Nuclear Power and its Fuel Cycle*, IAEA, Salzburg, 2nd-13th May, 1977, p. 3.
70. *Ibid.*

The provincial electrical power utilities have also undertaken a number of activities in recent years aimed at better informing and consulting the public on decisions concerning the siting of energy facilities and transmission lines. For example, the publicly-owned utility company, Ontario Hydro, has established a number of "citizen committees" and "working groups" with participants drawn from community agencies and interest groups, to assist it in the choice of future power plant sites and transmission corridors. These activities have evolved over time from a purely "informational" approach to more active efforts to involve citizens in "consultation" and "joint planning" study processes, although they stop short of delegating actual decision-making powers to the public.

The government nuclear R&D organisation (Atomic Energy of Canada Ltd., AECL) has, since the mid-1970s, also begun to assume a more active role in informing the public on nuclear-related activities. However, the difficulties encountered by AECL in overcoming public scepticism are not unlike those confronting similar nuclear R&D organisations and public utilities in many other countries. As one former AECL employee has noted, with no small sense of frustration:

> "On the one hand there is a vast amount of information, in the form of technical reports, which is available to the public. Opponents say that the very bulk of this literature and the professional jargon used... is a kind of "snow-job" which could bury damaging facts in a huge mass of paper. On the other hand, when AECL publishes information, well written in straightforward English, attractively illustrated and enclosed in coloured covers designed to catch the eye, there are cries of "glossy propaganda, printed at taxpayers' expense!"... whenever it tries to communicate with the public it is accused of wasting public money and indulging in self-serving propaganda, or just plain lying."[71]

Canada's nuclear regulatory agency, the Atomic Energy Control Board (AECB), has in recent years begun to take a more active role with respect to public information disclosure on nuclear safety and licensing matters. Proposed revisions to the Atomic Energy Control Act, still being considered by Parliament, stipulate: that public hearings be held before licenses are granted for the construction of any nuclear facility; that the AECB act as a source of "reliable, independent, public information on health, safety and environmental concerns"; and that all documents be publicly available unless specifically exempted by regulations.[72] As the past president of the AECB, Dr. A.T. Prince, recently commented, "we are in a new era of candor, an era in which the consequences of full disclosure are rarely, if ever, worse than the public discovery of duplicity or suppression of information".[73]

What lessons can be drawn from these different national experiences and approaches to promoting better public understanding with respect to nuclear energy and overall energy policy? First, it must be recognised that each of the approaches described reflects, to a certain degree, particular national structures, circumstances and traditions. The "study circle" mechanism, especially suited to the Swedish

---

71. Mawson, C.A., "Nuclear Power with Energy Conservation", *Alternatives,* Trent University, Peterborough, Ontario, Winter 1978, Vol. 7, No. 3, p. 31.
72. A final Parliamentary decision on revisions to the Atomic Energy Control Act is not expected until late 1979 or 1980.
73. Quoted in Dotto, L., "The Great Nuclear Debate", *Science Forum,* November-December, 1978, p. 46.

political context and tradition of seeking a social compromise, has only limited relevance for a country such as France, with its very different political traditions, hierarchical structures, and administrative practices. Similarly, a more highly-structured national information campaign approach, such as that adopted by Austria, would probably prove unsuitable for a much larger country with heterogeneous provincial energy interests and programmes such as Canada.

It is, moreover, necessary to distinguish between public information programmes undertaken by countries with substantial existing nuclear energy programmes (Sweden, Germany, France and Canada) and those of other countries (Austria and Denmark) where nuclear programmes were or are just getting underway. In the former, debate has tended to focus on issues of nuclear programme expansion, whereas in the latter, the emphasis has been on whether or even not to proceed with such programmes.[74]

In most cases, government approaches to the promotion of more informed public debate have been largely ad hoc. In Sweden, Austria and Denmark, government-initiated public information campaigns were terminated after only a short period of operation—often just as nuclear debates were culminating. Only in Germany, where the "Bürgerdialog" has been pursued, have such nuclear energy public information activities been maintained.

A number of important questions also arise as to the various purposes of government-initiated information programmes. One of the avowed objectives of such campaigns has been the promotion of more informed debate by improving levels of technical understanding: but it is extremely difficult to measure the real educational value of these programmes. Not only have the necessary pre- and post-analyses to permit such an assessment not been carried out, but opinion poll findings on public attitudes towards nuclear energy by no means suffice as a full measure of "public knowledge" in an area as complex and emotional as nuclear energy.

Moreover, the education of the public, in the narrow classical sense, does not appear to have been the single or primary aim of these activities. In many cases, the main motivations behind decisions to initiate public information campaigns appear to have been to defuse controversy, gain time and thus avoid having to take quick decisions that might have politically divisive repercussions. In some instances, the purpose of such campaigns has included the legitimising function of seeking to lend credibility and legitimacy to governmental decisions, past and pending.

This admixture of motives suggests the need to distinguish between activities aimed principally at "educating" the public and those directed at "informing" the public. With respect to the former, emphasis is often placed upon elucidating the technical and scientific content of the questions at issue. Here, national education systems, including those responsible for adult and continuing education, obviously have an essential role to play. Also important are the public education programmes and activities of intermediary institutions, such as trade unions, science councils and academies, religious groups, public interest organisations, as well as political parties.

However, education in such a narrow technical-intellectual sense is but one aspect of the problem of promoting more informed public debate in areas involving scientific and technological complexity. Not surprisingly, these national

---

74. This is reflected in public attitudes towards nuclear energy. In both Sweden and Germany, one presently finds the population fairly evenly split on the issue of nuclear programme expansion. In France and Canada, a clear majority of the population favour nuclear energy, whereas in Austria and Denmark, a majority of the general population is opposed.

experiences confirm that the general public is usually not particularly interested in knowing, for example, about the esoteric processes by which uranium oxide is converted to uranium hexafluoride, enriched, and subsequently converted to uranium dioxide for final fabrication into elements which in turn eventually fuel reactors which produce the power to make electricity to run the washing machine. They are generally more concerned with broader considerations such as reactor safety, spent fuel reprocessing and waste disposal, which they perceive as affecting their lives and well-being: that is to say, with the social, economic and political aspects of scientific and technological developments. Acceptance on the part of the public of certain risks appears to be heavily dependent both upon its government's ability to weigh all of these factors and on public trust in government institutions, their legitimacy and credibility.

Thus, the significance of efforts aimed at "informing" the public must be viewed from a broader perspective. In this sense, if information is to relate to the concerns perceived by the general public, it must seek to provide the elements necessary to understand and elucidate the implications of the political, economic and value choices with which people are confronted, and not just the narrower technological options available within a given course of action. Seen in this light, these first efforts chiefly directed to informing the general public on energy and nuclear-related policies certainly represent an encouraging attempt to achieve this aim. Though they may not have resulted in a higher level of citizen technical expertise in these matters, they have contributed toward making a large section of the general public more aware of the broader set of factors surrounding, for instance, the nuclear debate; a debate previously dominated by industrial interests and by exclusively scientific and technological considerations, capacities and expertise.

One must also add an important qualification when extrapolating from national experiences within the area of nuclear energy to other areas of scientific and technological controversy. This concerns the very special nature of nuclear energy and its place in contemporary societies. The issue of the civilian use of nuclear technology is fraught with enormous political, emotional and philosophical contention, not to mention its scientific and technical complexity. For many people, nuclear power has taken on a symbolic image; and images are often harder to change than realities.

The real and imagined hazards of nuclear power, the ways in which it has been managed and regulated, and the polarisation it has produced within the community of experts are all factors that have combined to engender mistrust and suspicion on the part of the general public.[75] It is therefore important also to look at other areas of scientific and technological controversy, in seeking the most appropriate ways for the promotion of broader public understanding.

3. Science Education Programmes

The idea of promoting broader public understanding on scientific and technological issues *in general,* not just on particular issues of current controversy such as nuclear power, has preoccupied science policy-makers in many countries, and most notably in the Netherlands and the United States. In these latter two countries, a number of different programmes have been initiated in recent years, which aim at the same general set of objectives:

— to promote increased public understanding of the technical, social and ethical aspects of scientific and technologically-related public policy issues;

---

75. Hohenemser, C., *et al.,* "The Distrust of Nuclear Power", *Science,* 196, 1st April, 1977, pp. 25-34.

— to facilitate greater citizen access to technical expertise by promoting closer interactions between citizens and scientific and technical experts; and
— to promote broader public participation in decision-making, especially on the part of those groups whose interests and views are (or have been) often unrepresented.

a) *Netherlands Information Programmes*

The genesis of initiatives undertaken in the Netherlands to promote broader public understanding in science and technology is to be found in the Government memorandum on science policy presented to the Parliament in 1974. This memorandum proposed several new approaches for the articulation of research policy and priorities, stressing the need for increased public participation in the formulation of research programmes and proposing the establishment of sectoral councils to advise ministers of research policies and plans.[76]

This memorandum also spoke of the need to better inform citizens on scientific and technologically-related matters and to promote closer interactions between research workers and the public. One of the first initiatives in this direction undertaken by the Ministry of Science in 1975 was to establish a programme aimed at "popularising" scientific research reports and findings for general public consumption. New courses and curricula were formed at the universities of Amsterdam and Utrecht to promote these activities.

At the same time, the Ministry established an advisory committee, comprised of scientists, government science-information officers and journalists, to discuss possible further measures to improve communication of information on science and technology to a larger public. This committee recommended that a central service for information on science should be created at the Royal Academy of Sciences.[77] This service would primarily serve a co-ordinating function, overseeing the information dissemination activities of existing university and government research institutes. Since the release of this report, the Academy has been performing this "service function" on a temporary basis, but a final government decision to formally institutionalise such a mechanism is still pending.

The government has also given consideration to proposals that it subsidise the information dissemination activities of non-governmental groups, especially those of foundations. A second committee was established to examine the arguments in favour of such a proposal and to review possible criteria for the selection of applications for such support. This committee concluded that public support for such groups was indeed justified,[78] and proposed a general "checklist" of points that should be considered in reviewing applications. However, as mentioned above, the government is still to decide on whether or not to provide funding for such non-governmental groups.

In the meantime, public debate in the Netherlands over government plans for the construction of three new nuclear-electric power plants by 1990 and for facilities for long-term storage of nuclear wastes has provided a new impetus to these efforts aimed at broadening levels of public understanding on scientific and technologically-related matters. In July 1978, the Minister of Economic Affairs announced the government's intention to promote a "broad societal discussion" on the nuclear issue before any such decisions are taken. He invited

---

76. Nelkin, D., *Technological Decisions and Democracy, op. cit.*, pp. 76-79. Parliament approved this memorandum in June, 1976.
77. *Nature*, Vol. 268, 18th August, 1977, p. 587.
78. Boeker, E., "Public Information on Science and Technology: The Dutch Case", *Science and Public Policy*, December, 1977, pp. 558-562.

interested parties to provide comments on how such a discussion should be carried out.

Shortly thereafter, the Parliamentary Permanent Committee for Nuclear Energy announced in response that it intended to prepare a report to the government, proposing a possible framework for the conduct of this discussion and also invited comments.[79] The government was expected to formally announce its own plans in mid-1979 for organising this "broad societal discussion", which, it is anticipated, will last two years. A final decision on the future of the Netherlands nuclear programme is expected by 1981. Thus, many of the same issues that have been raised with respect to general science information programmes—issues relating, for example, to government subsidies for non-governmental groups and criteria for their funding—will no doubt continue to receive increasing attention in the Netherlands in the coming years.

b) *US Science and Society Programme*

The US National Science Foundation (NSF) has undertaken one of the more ambitious national efforts aimed at promoting broader public understanding and expertise in the general area of science and technology. Under its recently-established (1977) Office of Science and Society, three separate programmes have been created to increase levels of understanding and information on scientific and technologically-related policy issues, encourage broader discussion and consideration of their ethical and social aspects, and promote the growth of greater citizen technical expertise.[80]

These initiatives reflect the Foundation's increased involvement in areas of social concern and represent a conscious effort to "increase knowledgeable participation of both scientists and non-scientists in the resolution of major public issues involving science and technology".[81]

The Foundation has adopted a decentralised approach towards the implementation of these programmes, stressing the importance of citizen self-initiative and education. It has specifically refrained from trying to convey any official "image" of science or position with respect to areas of political controversy. Under its "Public Understanding of Science" programme, it has sought to improve the content and processes of communication between scientific and non-scientific communities by supporting projects which seek to convey "balanced information that may illuminate the limitations of science and technology and the problems associated with their development, as well as their positive contributions to human welfare".[82] Examples include informal public education programmes conducted via the mass media and by science museums, as well as numerous state and local forum discussions, conferences and workshops.[83]

NSF has also sought to promote increased public awareness and discussion of the ethical and value implications of scientific research and its applications

---

79. See Interim Report entitled "Maatschappelijk discussie over de toepassing van Kernenergie voor electriciteits opwekking", 15100, No. 14, 30th January, 1979.

80. These programmes include: Public Understanding of Science (PUOS), Ethics and Values in Science and Technology (EVIST), and Science for Citizens (SFC). Total budget for these activities in 1977 was about $5.4 million.

81. National Science Foundation, Office of Science and Society, "Programme Announcement", 10th March, 1978, NSF, Washington, D.C., 1978.

82. Blanpied, W.A., *et al.*, "Science, Technology and the US Public: Three Experimental Programmes", paper prepared for the UNESCO-sponsored MINESPOL II Conference, January, 1978 (draft), p. 21.

83. Activities included support for a weekly, hour-long nationally-televised programme, *NOVA;* for the Association of Science and Technology Centers representing more than 60 science museums; and for special forums and conferences.

by supporting interdisciplinary research, conferences and other activities under its programme on "Ethics and Values in Science and Technology". In addition, under its "Science for Citizens" programme, the Foundation has focused on developing mechanisms to provide scientific and technical information that will permit citizens and the organisations they represent to participate "more effectively" in decision-making processes. The current mechanisms being experimented with include a limited number of so-called "Public Service Science Residencies" and "Internships" awarded to individual scientists and engineers, to permit them to undertake projects in which their scientific and technical expertise is made available to particular citizens' organisations.[84]

Each of these three experimental programmes has been underway since early 1977, and it is likely that, with increased experience, the approach taken by the Office of Science and Society may be further broadened to include new types of social mechanisms. For example, during 1978 the Foundation has been considering establishing a network of "public service science centres", to provide technical expertise and serve the information needs and interests of selected national regions and communities.[85]

Most of these NSF initiatives for the promotion of broader public understanding and citizen expertise have received considerable support from the government, the scientific community, and individual "public interest" and other citizens' groups and organisations. However, some degree of controversy has developed over the "Science for Citizens" programme.[86]

This controversy concerns whether or not the Foundation should provide direct financial assistance to citizens' groups, some of which are or might become directly involved in citizen litigation or other forms of direct intervention in government decision-making and administrative proceedings. After considerable Congressional debate and analysis, the Foundation was directed to avoid supporting any proposals that might eventually lead to such forms of direct intervention.[87] However, this controversy over citizen group financing has by no means been finally resolved or laid to rest. As in other countries where this same issue has been raised, it reflects an on-going debate between opposing viewpoints over not only the functions of government but over the very concept of the "public interest"; it is a debate to which we will return later in this study.

4. Science, Technology and the Media

The role of the news media as an important and essential channel for the promotion of broader public understanding on scientific and technologically-related matters has been the subject of increasing analysis and debate in many

---

84. Science and engineering graduate and undergraduate students are eligible to receive the Internships; the granting of a total of 15 to 25 Residencies and Internships were planned for 1978. Additional mechanisms include financial grants for forums, conferences and workshops.
85. Blanpied, W.A., *et al., op. cit.,* pp. 27-28.
86. US Congress, House of Representatives, Committee on Science and Technology, *Authorisation Report for FY 1978,* Report No. 95-98, GPO, Washington D.C., 18th March, 1977, pp. 20-23.
87. This controversy is especially highlighted in the different positions taken in the United States Congress. In general, the United States Senate favoured providing assistance to "groups which serve important public purposes". Members of the United States House of Representatives (Committee on Science and Technology) believed on the other hand, that "The NSF should remain as far away as possible from direct assistance to citizens' groups". *Ibid.*

countries over the last decade.[88] The media have brought government decision-making processes under closer public scrutiny and have provided citizens with more timely information on all aspects of daily life than ever before. The power of the media in shaping and catalysing public opinion, especially in areas of major technological controversy, is undeniable. However, media coverage of issues related to science and technology is often uneven, incomplete and highly selective.

Problems related to environmental pollution, nuclear power and energy policy, health and, more recently, biomedicine are treated in considerable depth, whereas other issues of new or emerging public concern often receive only minimal or passing coverage. One constraint on science journalists, especially in the written media, relates of course to the competition they face for space in newspapers and magazines. Although national journalistic traditions vary widely both between and within countries, a number of different study findings confirm the generally small amount of space allocated in most daily newspapers to articles concerning science and technology. For example, one recent study of 11 daily newspapers in Switzerland indicates that articles on science and technology comprise an average of less than 1.5 per cent of news space.[89]

More serious, however, are the problems often encountered by science writers in staying abreast of scientific developments occurring in many different fields simultaneously. Many younger journalists today are better educated and have better science backgrounds than their predecessors. They are also more sceptical of science than older science writers. However, given the economic pressures on daily newspapers, few can afford to hire a large number of specialist science writers. In addition, a more endemic problem relates to misunderstandings that often develop between the media and the scientific community. As one report of the British Association for the Advancement of Science points out, this is a mutual problem:

> "Some scientists, failing to recognise the aims of the journalists and broadcasters, confuse the media with the scientific "literature". It is also true that in seeking, rightly, to publish the most recent findings, the media do not always convey how such findings are inevitably scrutinised and assessed before they are accepted as reliable scientifically."[90]

That science writers affect public attitudes towards science and are, in turn, affected to some degree by these same attitudes appears to be increasingly evident. According to one study undertaken in the United States on the relationship between science writers and public attitudes towards science, there has been a trend away from "short, straight news pieces to longer interpretative articles, focused on people-oriented issues". It was found that over the last decade science writers often presented science from a more unfavourable viewpoint than in 1965. Over 80 per cent of the writers surveyed in this study reported "writing articles in 1972 that could be interpreted by readers as unfavourable to science".[91]

---

88. Dubas, O., and Martel, L., *Science, Mass Media and the Public,* 3 Vols., Ministry of State for Science and Technology, Ottawa, 1973, 1975 and 1977; Boss, J.F., and Kapferer, J.N., "Le public et la vulgarisation scientifique et technique", *Le progrès scientifique,* No. 190, Paris, September, 1977, pp. 3-44; Garrigues, C., "De la communication de masse à la communication de groupe", *Futuribles,* No. 16, Paris, July-August, 1978, pp. 467-478.

89. Daetwyler, J.J., "L'information scientifique dans les quotidiens suisses", in *24 Heures,* Lausanne, 13th June, 1976.

90. British Association for the Advancement of Science, *Science and the Media,* BAAS, London, 1976.

91. Friedman, S., "Science Writers and Public Attitudes Toward Science", in *Science in the Newspaper,* an occasional paper of the American Association for the Advancement of Science, AAAS, Washington, 1974.

In a number of countries, special efforts have been made to provide science journalists with broader opportunities for familiarising themselves with information on the content of government-initiated scientific and technological programmes and projects. Special technical briefings, government internships for younger journalists, and programmes aimed at promoting closer contacts between science writers and scientific and technical specialists represent three such approaches. In addition, as we have seen in the case of the Swedish study circle experiment and in the conduct of energy information campaigns in Germany and Austria, greater recognition has been given to the need to improve levels of technical understanding of science journalists to assist them in fulfilling their functions as an important link in the process of transferring information from government and the scientific community to the public-at-large.

In Denmark, the Netherlands and the United States, university curricula and courses have been expanded for the teaching of science writing, and increased emphasis has been placed on efforts aimed at promoting the "popularisation" of scientific and technical research findings. Nevertheless, in most countries such efforts are still largely ad hoc and generally restricted to highly specific areas of science and technology. Moreover, there is a general lack of systematic knowledge on public attitudes towards science and how research is conducted. Even less is known on how citizens "learn" in areas of scientific and technological complexity. Despite the increasing influence of television and other audio-visual methods of communication, the public still receives most of its information on science and technology from newspapers and magazines; rarely does it come from formal educational processes.[92] And yet, very few international comparative studies have been carried out to date to ascertain the most effective media methods for improving levels of public understanding and for enhancing the capabilities of media specialists in the performance of this function.

Government efforts to better inform the public on the nature, scope and content of decision-making are rapidly developing in a number of countries. The liberalisation of some laws, regulations and policies governing the disclosure of government information has, in principle, provided degrees of public access to the conduct and substance of government business heretofore perhaps only dreamed of by civil libertarians. Increasing acceptance of the public's "right" to know what is going on in government has led, in a growing number of countries, to more open and explicit attempts to advise citizens on decision-making procedures and participatory opportunities.

In most areas of public policy, a large and diverse number of intermediary institutions exist which serve, among other things, as conduits or channels between government and its citizens. They are able not only to secure access to information in the possession of government and transfer it to concerned citizens but, in reverse, to provide government with information on citizen needs and demands. However, one does not find the same rich diversity of intermediary institutions within many scientific and technologically-related policy areas. Scientific and technical information, when and if it is made available at all, oftentimes passes directly to citizens from government or the scientific community. Moreover, given the technical complexity of much scientific information, citizens are often hard put to assimilate such information, let alone fully appreciate its relevance and manifold implications.

---

92. *Ibid.*

The news media have begun to devote increasing attention to scientific and technical policy matters, and efforts in this area are to be encouraged. Nevertheless, the general scarcity of intermediary institutions, coupled with the unresponsiveness or inability of many existing institutions, such as political parties, to provide adequate forums or channels for the expression of public concerns and needs in areas related to science and technology, are of increasing concern to some people. This suggests that greater attention should be focused on developing new or alternative institutional mechanisms and approaches to meet public needs and demands for information on scientific and technologically-related policy matters.

One of the more fascinating developments over the last half-decade is growing government involvement in the so-called "public information campaign business". This is especially, but not exclusively, evident in relation to areas of scientific and technological controversy. A major motivation behind such initiatives, as we have seen in the area of nuclear energy, has been to provide for more informed public debate by improving levels of public understanding on the scientific and technical content of associated issues. This has obviously not been the sole motive of governments, and the very admixture of purposes has in fact led to unclear and sometimes confusing results. It has also led, in a number of countries, to increased conflict.

Government public information campaigns have also provided an ideal target for those citizens and activist groups seeking to give broader public currency to their own particular views. Experiences with government-initiated nuclear energy information campaigns and committees provide a number of examples of this, and also attest to the difficulty of trying to prematurely narrow the scope of discussion, without fully grasping the roots of the controversy.

Efforts to promote more informed public debate, whether in the field of nuclear energy or in other areas of scientific and technological controversy, require that the strategies adopted fit the objectives sought. If the aim of an information campaign is to defuse political controversy, it is unlikely that technical facts and forms of argumentation are the most appropriate tools for the task. Similarly, efforts to achieve social consensus are severely impeded if there is no consensus in the first place on what is in disagreement.

Information is only as credible as its source. One of the major, persistent difficulties encountered by government institutions involved in the conduct of public information campaigns has been that, because their own institutional credibility is often questioned, so, too, is the information they provide. But the problem is not solely one of institutional credibility; it also relates to the particular image of science or of technology that one tries to present or seeks to impose.

These problems suggest the importance and value of efforts aimed at encouraging decentralised approaches to citizen self-education and public opinion formation in areas of scientific and technological controversy. In fact some tendencies in this direction can be seen in the German "Bürgerdialog" and the United States programmes under the aegis of the National Science Foundation. Equally significant and important are current trends in the direction of promoting the development of more pluralistic sources of technical information and expertise available to the general public.

Timing also plays a key role in most efforts aimed at promoting broader public understanding. As these experiences suggest, trying to implement a public information campaign once public opinion is already polarised and an issue highly politicised presents an almost insurmountable task. In some areas, such as those associated with energy policy, this has been difficult to avoid, owing in part to the suddenness with which energy matters have come to dominate national concerns. Nevertheless, many would agree that it does not take a

Delphic oracle today to foresee some of the emerging scientific and technologically-related public controversies of tomorrow. Efforts to promote early public discussion and understanding do not amount merely to practicing political preventive medicine, but to trying to ensure that future debates will provide the necessary basis for truly democratic decisions.

Finally, and most importantly, one must place the problem of public information and understanding (or the alleged lack thereof) in the proper perspective. For, it would be too simplistic to conclude that public opposition or protest in areas of scientific and technologically-related controversy, such as nuclear energy, can be attributed solely or even mainly to insufficient public knowledge and understanding. Many of those who oppose certain scientific or technological programmes are extremely well-informed as to their details and associated risks. Nor can attitudes of public scepticism be explained by one reason alone. Often they result from a broad panoply of public concerns and misgivings about the social goals to be pursued, the protective measures to be taken, and the way costs and benefits are to be distributed in society-at-large.

Moreover, access to information without effective means for its use is like possessing a lever without a fulcrum. Public demands for information are, therefore, closely associated with demands for broader and more direct opportunities for public participation in government decision-making. Governments have sought to respond with a number of measures aimed at facilitating communication between citizens and policy-makers.

III

INFORMING POLICY-MAKERS

The emergence of increasing public demand for more active and direct forms of participation in decision-making on issues related to science and technology has elicited considerable and varied government response. Many governments have come to recognise the need to develop more effective methods and mechanisms for assessing public opinion and for consulting the general public on specific areas of major policy concern. In some countries, more active approaches have been taken to increase opportunities by which members of the public can express their needs and concerns directly to policy makers. In a few countries, this has meant granting greater public access to government decision-making processes and forums.

Membership on government advisory bodies has been expanded to include broader representation of various public interests, and legislative "hearing" procedures have been adapted to ensure a broader presentation of competing public views. Increasing government recourse to commisions of inquiry, as a mechanism for informing the general public and policy-makers alike, is especially evident in areas of major controversy involving large-scale technological projects where local and national interests conflict. In addition, a variety of special ad hoc mechanisms have been developed to better assess public attitudes and preferences.

Elected officials have been especially receptive to approaches aimed at securing timely and relevant information on public desires and needs. This is quite understandable given their preoccupation with re-election. Opening up new two-way channels of information between elected representatives and the public is therefore considered beneficial and, under circumstances of major national controversy, both important and necessary. However, politicians are often more reserved and cautious in their support for public participation when it implies opening up new channels for the exercise of increased public influence or power in government decision-making. They are especially sensitive about attempts by government ministries and administrators to promote participation and channel public support for programmes or policies that may not be consistent with the mandates or guidelines laid down by legislative bodies. They are even more reserved in their judgments when they perceive efforts at direct public participation as threatening their traditional bases of political power or bringing into question some of the basic principles of representative systems of government.

For many individuals and citizen groups, however, public participation means more than just being consulted occasionally or having an opportunity to express their views and opinions. They consider government initiatives to facilitate processes by which the public can inform policy-makers as being insufficient. They want to be more directly involved in the exercise of decision-making power; to influence not only the substance of government decisions but the processes by which those decisions are made.

A. ADVISORY BODIES

Most governments have established rather elaborate advisory structures by which representatives of different groups and interests can express their views on given public policy issues. Such advisory boards, commissions, and councils are as diverse and numerous as the matters with which they deal. In the United States, for example, more than 1,267 advisory boards with a total of 22,256 members have been constituted as of 1975 to advise 45 Federal government agencies. In the Australian State of New South Wales, there are no less than 681, of which 551 have a government/non-government mix in their membership. And, in Germany, the Ministry for Science and Technology (BMFT), which uses more technical advisory commissions and expert consultants than most other German government ministries, employed 927 technical advisors from various fields of activity in 1975.

Government recourse to such advisory mechanisms has traditionally been predicated upon the assumption that the public interest is best served if those whose interests are or might be directly affected by government decisions have an opportunity of shaping those decisions before they are finalised. Such mechanisms have also served as an important means for enhancing governmental technical expertise.

It is, therefore, perhaps not surprising that those groups who feel they have the most to gain (or lose) immediately as a result of certain government decisions or regulations are well-represented on such advisory bodies. For instance, nearly 50 per cent of the total membership of the US Federal agency advisory boards is comprised of industrial representatives, with only 7 per cent coming from consumer and environmental groups.[93] One also finds a "massive over-representation of producers' groups... and gross under-representation of all other groups" on the advisory committees attached to Australian Government Departments.[94] While, at the German BMFT, studies have shown a similar representational bias in favour of researchers (55 per cent) and industrialists (25 per cent) on expert commissions responsible for advising the Ministry on science policy matters.[95]

Such limitation in the composition of governmental advisory bodies is often reinforced by the fact that many of them do not provide their members with travel or sitting fees, and thus discourage participation of many people who cannot afford time off from work or whose organisations cannot reimburse

---

93. Sullivan, J., "Public Participation in Federal Energy Advisory Committees", draft report to the US National Council for the Public Assessment of Technology, Washington, 1976 (mimeo).

94. "Producers' groups" are defined as trade and business associations, trade unions, professional and primary producers' associations. This is partly explained by the greater use made by the then Departments of Agriculture, Labour and Immigration and Manufacturing Industry of advisory committees, consisting largely of their industry (including trade union) clients. See: Matthews, T., "Interest Group Access to the Australian Government Bureaucracy", in *Report of the Royal Commission on Australian Government Administration*, Appendix, Vol. 2, AGPS, Canberra, 1976, pp. 340-341.

95. See: "Forschungs-und Technologiepolitik; Gestaltung wirtschaftlichen Wandels durch Innovation", *Bericht der Kommission für wirtschaftlichen und sozialen Wandel*, Bonn, 1976, p. 473.

them. There are, of course, some exceptions to this pattern, as for example in the case of the US Federal Food and Drug Administration (FDA), which provides for paid consumer representatives on its advisory panels.[96]

Representational bias is however but one of the limitations of this form of participatory mechanism. Often such advisory bodies perform only a "window-dressing" function, with their meetings and reports manipulated by agency personnel, to foster the appearance of public support for predetermined policies and decisions. The use of such technical expertise can thus serve a purely legitimising function, especially when the minutes of advisory board meetings are kept confidential or the general public is denied access to such meetings. In other cases, broadly representative government advisory groups are established, but insufficiently funded or denied access to government information and research findings, thus limiting their potential scope of effectiveness.[97] Moreover, even when advisory bodies produce independent recommendations, government ministries or agencies are not obliged to implement them. Nevertheless, consistent agency disregard for such recommendations is rare, owing partly to the fact that government agencies are dependent upon the co-operation of those groups whose interests are represented on advisory boards for effective policy implementation. A number of efforts have been made in recent years to open up government advisory bodies to broader public participation and representation.

For example the Danish Energy Council (*Energirådet*), established in 1976 to advise the Ministry of Industry and Commerce, is comprised of representatives from major institutionalised interest groups, including delegates from trade unions, industry, the energy utilities and the Consumer Advisory Council.[98] Similarly, the French advisory information council on nuclear energy includes representatives from industry, the scientific community and ecology groups. In the United States, the Federal Advisory Committee Act (1973) requires that most government part-time advisory committees meet in public and that membership lists be printed in the *Federal Register*.

The US Commission for the Protection of Human Subjects of Biological and Behavioural Research, which we refer to below, is constituted as both an expert and representative group to reflect the differences inherent in conflicts over biomedical research;[99] while, at the US National Institutes of Health (NIH), informed laymen have been included on advisory councils responsible for the allocation of research funds on the basis of NIH priorities and social considerations.[100] Among the 18 members of the National Heart and Lung Advisory Council are five members from the general public, who are "leaders in the fields of fundamental or medical science or in public affairs".[101]

Despite these and other significant reforms in government advisory processes, certain groups whose direct interests are physically influenced or otherwise impinged upon by government decisions often remain un- or under-represented. For instance, in the example of the German BMFT advisory structure, DGB

---

96. *Federal Register*, 41, 26th November, 1976, p. 52148.
97. For example, see: McCullum, H. and K., and Olthuis, J., *Moratorium: Justice, Energy and the Native People*, The Anglican Book Centre, Toronto, 1977, p. 23.
98. The Council does not, however, include representatives from environmentalist groups or the OOA (Organisation for Information on Nuclear Power).
99. Department of Health, Education and Welfare, "Protection of Human Subjects, Fetuses, Pregnant Women, and in Vitro Fertilisation", *Federal Register*, 40, 154, 8th August, 1975, p. 33526. See below, p. 77.
100. Nelkin D., "Threats and Promises: Negotiating the Control of Research", in *Daedalus*, Spring 1978, pp. 191-206.
101. Gustafson, T., "The Controversy over Peer Review ", *Science*, 190, 12th December, 1975, pp. 1060-1066.

(trade union) representatives comprise less than 5 per cent of the total advisors to the Ministry; a relatively small percentage in view of the influence of technological change on employment and working conditions. Nevertheless, some of the more innovative approaches to adapting advisory procedures to the participatory needs of heretofore unrepresented groups are to be found at the project implementation stage of decision-making. In Canada, for example, where plans for the construction of an Alaska Highway (Alcan) natural gas pipeline have been the centre of heated public debate, the pipeline contractor has proposed that four Yukon Indians sit on the 11-member advisory management board to participate in decision-making on the pipeline project implementation.[102]

It is extremely difficult to assess the real effectiveness of these advisory processes as participatory mechanisms. One possible measure of effectiveness might be whether advisory bodies demonstrably altered the decisions that would otherwise have been taken. But evidence of this kind is difficult to obtain. In many cases, advisory body findings and recommendations are but one set of information influencing decisions. Decision-making authority is itself often dispersed, and even when an agency head or minister accepts such advice, his own decision may be overturned by legislative bodies or administrative appeal courts.

It has been suggested that some governmental advisory mechanisms, given their traditional representational biases and procedural shortcomings, pose risks that policies and decisions will heavily reflect the interests of their predominant members and not always those of society-at-large. In fact, this is one of the main criticisms raised by people sceptical about governmental technocratic tendencies to consider problems in narrow technical terms and in terms of only the most readily identifiable group interests. And yet, clearly such mechanisms can serve an important function. Not all government decisions can be formulated in the community forum; advisory processes allow for a more focused consideration of alternative strategies and information needs before decisions are exposed for public debate. They also provide an important source of expertise, not always available within government itself. Efforts to ensure a broader representation of affected interests on advisory boards and more timely access to information on board proceedings can help to reduce levels of public scepticism.

Advisory mechanisms are, nevertheless, only of limited utility as a mechanism for broad-scale public participation. Most citizens lack the time, expertise or interest required to participate fully in such advisory processes. Nor are such processes really geared to resolving issues of major public controversy. Other participatory mechanisms are perhaps more suited to public demand and needs for participatory involvement.

B. LEGISLATIVE HEARINGS

One of the most frequent responses to pressures for greater public input in policy-making has been to expand public hearing processes. Public hearings are a generic mechanism employed in some countries at national, state and local decision-making levels by some legislative and parliamentary bodies as well as by most administrative agencies, tribunals and regulatory boards. The most rapid expansion of such public hearing processes over the last decade relates to these latter administrative bodies, which we discuss later. Here, we focus solely upon legislative and parliamentary structures.

---

102. *Toronto Star,* Toronto, 18th July, 1977.

Public hearings can serve several functions.[103] Sometimes they provide legislators with an opportunity of simply informing citizens about the nature of a project or proposed legislation. At other times, they serve a co-optation function to defuse conflict and avoid litigation by allowing irate citizens to complain and "let off steam", without necessarily altering the proposed project or policy. When public hearings are required by law or tradition, but the issues or decisions are themselves non-controversial, they often take on a purely ritualistic function—for example, when policy-makers are committed to ascertaining public views and needs as a basis for legislative action. Rarely, however, do public hearings serve this ideal function.

One major problem of the public hearing is that the views presented do not always represent the full range of individuals affected by a proposed project or policy. Often, they reflect only the views of identifiable interest groups which are part of the established clientele of policy-makers and attuned to their special information needs and hearings approaches. Individual citizens or non-established groups are often ill-informed about public hearings and, for a variety of reasons, ill-prepared to present their views effectively. While they may be knowledgeable about how they feel, individuals often lack an overall perspective as to the implications of issues and sometimes have little more than a superficial understanding of technical details.[104] Other more practical factors such as the scheduling of public hearings during working hours, costs of attendance, and competing demands on the time required to undertake research and prepare presentations also mitigate against broad public representation in such hearings. Policy-maker motivations also directly influence public hearing processes. To no small degree, it is they and their legislative staff who determine what viewpoints will and will not be considered in public hearings. However, abuse of public hearings to promote the special interests of policy-makers is rare.[105]

In no other national legislative or parliamentary body is more extensive recourse made to public hearings than in the United States. Congressional hearings serve as the principal mechanism by which legislators collectively assess public views and interests related to legislative proposals. Moreover, the bicameral nature of the United States Congress, with its separate Senate and House committee structures, provides a system of checks and balances as well as sometimes competitive forums for public hearings. One also finds, at the same time, that public demand for more direct forms of participation in United States legislative decision-making has stimulated a number of related Congressional reforms.

One of these is exemplified by attempts to increase Congressional competency to deal with problems of scientific and technological complexity. Congressional committee staffs have been expanded, increasing recourse has been

---

103. See: Heberlein, T.A., "Some Observations on Alternative Mechanisms for Public Involvement: The Hearing, Public Opinion Poll, the Workshop and the Quasi-Experiment", *Natural Resources Journal* (USA), 16th January, 1976, pp. 197-212.
104. See : Grima, A.P. and Wilson-Hodges, C., "Regulation of Great Lakes Water Levels: The Public Speaks Out", *Journal of Great Lakes Research*, Vol. 3, No. 3-4, December, 1977, pp. 240-257.
105. According to one critical study of the former US Joint Committee on Atomic Energy (JCAE), this legislative committee consistently prevented, during its 30-year history: "fair and reasoned consideration of the controversies surrounding nuclear power by failing to respond to proposals by other members of Congress, by failing to give critics of the nuclear program a forum and a hearing, by failing to explore issues that raise doubts about aspects of the nuclear program, and by using its powers as a joint committee to gain procedural advantages for legislation it favoured". See: Belford, T., *Stacking the Deck*, Common Cause, Washington, D.C., 1976.

made to technical specialists in the Congressional Research Service and, of course, an Office of Technology Assessment has been created. In addition, the General Accounting Office (GAO), which serves as the Congress' watchdog agency over the Executive Branch, has been revamped and staffed with more scientists and engineers to enhance capabilities for technical programme evaluations.[106] One of the aims of these reforms has certainly been to redress an increasingly apparent imbalance between Executive and Legislative technical capabilities. An equally significant reason has been to try to reassert the power of representative democracy; to provide a countervailing force to public pressures for more direct democracy, which are understandably perceived as threatening by many members of Congress.

One manifestation of this concern can be seen in attempts to make Congressional procedures more open and accessible to the general public. Committee meetings have been televised, and closed-circuit television systems between Washington and local communities have been experimented with to allow local citizens to participate in public hearings without having to come to Washington. Many of these reform measures have also begun to filter down to US state legislative assemblies, which find themselves increasingly confronted with complex policy matters requiring scientific and technical assistance and advice.

This long American tradition of dependency on public hearings as a primary participatory mechanism for informing legislators is not to be found in most other national legislative and parliamentary bodies. For example, in Denmark, as in other Scandinavian countries, public hearings by parliamentary bodies do not exist. Other more informal channels are available to interest groups and individuals, which are a reflection of more than anything else, the political culture differences between the United States and these countries. Similarly, in both France and Germany, public hearing processes are generally restricted to administrative and regulatory decision-making.

In those countries with ministerial forms of government, such as Australia, Canada and the United Kingdom, parliamentary public hearings serve a more limited function. This is largely due to the fact that policy decisions are traditionally accepted as matters for the executive organ of government, not parliament. In each of these countries, a variety of select and standing parliamentary committees exist which occasionally hold public hearings, but they generally do not have the same degree of impact as the hearing processes carried out in the United States, at least where matters of science and technology are concerned. One important exception is the British House of Commons Select Committee on Science and Technology, which has held a number of important hearing into, for example, Lord Rothschild's proposals for reorganising research council funding and into Britain's choice of nuclear reactors. More recently, it has been involved in reviewing the implications of current research using recombinant DNA techniques. In Australia, a Senate Standing Committee on Science and the Environment was established in March 1976.

A more general response in these latter countries has been to rely on extra-parliamentary institutions as a primary means for informing policy-makers. These include such mechanisms as Royal Commissions, major inquiry commissions, administrative tribunals and public inquiries.

---

106. "GAO Beefs up Staff with Scientists, Engineers", *Chemical and Engineering News,* 22nd November, 1976, pp. 14-15.

C. COMMISSIONS OF INQUIRY

Commissions of inquiry are one of the oldest institutions of public consultation in parliamentary systems of government.[107] They have a long tradition of providing considered advice to government on a broad range of issues. They have sometimes functioned as impartial investigative bodies and at other times as a means for simply dissipating controversy or sweeping contentious issues under the rug. In more recent years, they have taken on a new and broader function: that of opening up issues to public discussion and of providing more representative opinion to decision-makers, so as to properly inform them of the needs and wishes of the people.[108]

Commissions of inquiry take a number of forms. There are Royal Commissions, formally appointed by authority of the Crown for a fixed duration, to examine and report on specific policy matters or proposed projects. Examples of these include the Ontario (Canada) Royal Commission on Electric Power Planning and the Swedish Royal Commission on Energy Policy.[109] Other Royal Commissions are established for an indefinite duration to provide continuing advice to government on more generic policy concerns. These include the standing Royal Commission on Environmental Pollution in the United Kingdom and the Royal Commission on Australian Government Administration.[110]. Royal Commissions can be appointed at Federal as well as Provincial government levels.

A second form are so-called National Public Inquiries, established either by Cabinet or by individual government ministries under a variety of legislative acts, such as those concerned with environmental protection and physical planning. These are often concerned with specific projects or plans that have considerable national policy significance. Recent examples include the Australian Ranger Uranium Environmental Inquiry (Fox Inquiry) and the Canadian Mackenzie Valley Pipeline Inquiry (Berger Inquiry).[111]

A third form are Public Local Inquiries, established by national or provincial government ministries, to examine issues of local concern related to proposed local projects. The Windscale Public Inquiry (Parker Inquiry) into plans for the construction of a thermal oxide reprocessing plant (THORP) for spent nuclear fuel is one example of this form.[112] However, as we shall see, the theoretical distinction between local and national concerns and impacts is often a difficult one to maintain in practice.

Royal Commissions and national and local public inquiries generally enjoy the same powers of enforcing the attendance of witnesses and examining them

---

107. The earliest example of a Royal Commission dates from 1403, when an investigation was undertaken into the Bedlam (UK) mental hospital, where one Peter Taverner was alleged to have stolen articles and money from the institution. See: *Australian Law Journal*, Vol. 35, 30th November, 1961, p. 271.
108. See: Law Reform Commission of Canada, *Commissions of Inquiry*, Administrative Law Working Paper 17, LRCC, Ottawa, 1977.
109. See: *A Race Against Time - Interim Report on Nuclear Power in Ontario*, (*Porter Report*), Royal Commission on Electric Power Planning, Ontario, September 1978; and *Report of the Commission on Energy Policy*, Stockholm, March 1978.
110. See: *Report of the Standing Royal Commission on Environmental Pollution*, Report No. 6, HMSO, London, 1977, and *Report of the Royal Commission on Australian Government Administration*, AGPS, Canberra, 1976.
111. See: *Report of the Ranger Uranium Environmental Inquiry* (*Fox Report*), Final Report, AGPS, Canberra, 1977, and *Report of the Mackenzie Valley Pipeline Inquiry* (*Berger Report*), Vol. 1 and 2, Supply and Services Canada, Ottawa, 1977-1978.
112. *The Windscale Inquiry* (*Parker Report*), HMSO, London, 1978.

under oath, of compelling the production of documents, and of punishing persons guilty of contempt. Their terms of reference are usually quite broad, and considerable discretion is granted to commissioners for defining the scope of inquiry. Commissions may be comprised of one or several members and there has been a tendency to appoint judges as commissioners, as in the case of the Fox, Berger and Parker inquiries cited above. Finally, although commission recommendations are not binding on government, examples are rare where they have been completely disregarded.

Public inquiry procedures also differ, however, in other important respects from those of Royal Commissions. They are essentially semi-judicial in nature, subject to considerable flexibility, and entirely "open", in that all evidence has to be given in public. To no small degree, the functioning of such public inquiries is often a reflection of the personal style of the individual appointed to conduct the inquiry.

Government recourse to commissions of inquiry as a means for coping with scientific and technologically-related public controversies is especially marked in recent years. This is evident in such areas as uranium mining, spent nuclear fuel reprocessing and storage, natural gas pipeline construction, and electric power planning. The reasons for this are two-fold.

First, all of these issues involve levels of scientific and technical complexity and uncertainty that require considerable detailed, thorough and time-consuming analysis. Parliamentary bodies, it is generally recognised, are by necessity preoccupied with problems of immediate political concern. Legislators often lack the time, staff and resources required for sustained and detailed inquiry into such complex matters. Moreover, as noted in the previous chapter, political opinion on many of these issues is often highly polarised, producing cleavages running through most major political parties. Recourse to advisory commissions is, therefore, viewed as one means of depoliticising such issues, and of securing needed technical advice, while at the same time reserving the ultimate parliamentary prerogative for debating and deciding policy questions in the light of Commission findings and recommendations.

Second, commissions of inquiry complement traditional means at the disposal of policy-makers for assessing the full range of public opinion on given issues. Moreover, some issues more directly affect certain minority group interests—for example, those of the Aboriginals in the case of Australian uranium mining or of Inuit (Eskimos) and Indians in the case of northern gas pipeline construction in Canada—which have heretofore been under-represented in most government policy-making forums.[113] The Commission mechanism is thus also often considered an important means for promoting minority group involvement and for trying to reconcile these and other competing group interests with those of the general public.

If commissions of inquiry have increasingly come to serve the function of encouraging broader public discussion and participation in government decision-making, as well as of promoting more informed public debate on scientific and technologically-related issues, it is important that we examine in some detail how they have actually functioned and whether or not they have had the means to adequately perform these functions. We limit ourselves to consideration of two examples: the Windscale Inquiry and the Mackenzie Valley Pipeline Inquiry.

---

113. Amongst other measures aimed at improving access by Aboriginals to government decision-makers, the Australian Commonwealth Government established in December 1977 the National Aboriginal Conference and, in July 1978, the Council for Aboriginal Development.

1. The Windscale Inquiry

The Inquiry into plans by the nationally-owned British Nuclear Fuels Limited (BNFL) to build a $1.2 billion thermal oxide reprocessing plant (THORP) for spent nuclear fuel at West Cumbria on a site called Windscale began on 14th June, 1977. It lasted 100 days. It has been variously described as a "total consequential impact analysis", "an epic example to the world", and "an international test case".[114]

A planning application of this magnitude with such local and international implications would normally have been dealt with at a formal Planning Inquiry. A number of factors served to increase public awareness of and support for the decision that one should be held: intensive lobbying on the part of environmentalist and anti-nuclear groups; public fears of Britain becoming the "world's nuclear dustbin"; internal government disagreement and confusion over nuclear policy in general and THORP in particular; a political storm over BNFL's delay in reporting a radioactive leak at one of its solid waste silos; media pressure for broader public debate; local government misgivings over its prior approval for construction plans; and a new US policy position on spent nuclear fuel transfers.

Had the Government chosen to hold such an inquiry in 1975, when detailed plans were first being formulated and discussed within government circles, it is likely that public opposition would have been far less significant then than it was at the eventual Inquiry. As it was, the decision to appoint a Local Public Inquiry was made at the last minute, at the very height of political controversy.

The Inquiry was formally established on 7th March 1977, by the Department of the Environment, under Town and Country Planning legislation statutes which did not require that terms of reference be specified. However, Environment Minister Peter Shore indicated that, upon receipt of the Inquiry findings and recommendations, he would consider "all relevant aspects of the proposed development", including its "implications for the safety of the public and for other aspects of the national security", as well as for the "environment of the construction and operation" of the facilities in accordance with existing nuclear legislation.[115]

Mr. Justice Parker, a High Court Judge from the commercial circuit, was named Inquiry Chairman and granted broad discretionary authority to determine the course and scope of the Inquiry.[116] He was widely respected for his careful attention to technical detail and known for his expeditious handling of an earlier Court of Inquiry into the British Flixborough chemical plant disaster. From the very outset, Mr. Justice Parker indicated that he viewed the Inquiry as a general debate between proponents and opponents of nuclear power. However, he forcefully maintained that its purpose was to test the *facts* of government policy, not its merits. "The proper place for determining policy is in Parliament, not here", he noted at the preliminary meeting on 17th May 1977, to discuss procedural matters. In addition, he noted that, as is normal with such inquiries,

---

114. For a detailed account of the background and conduct of the Windscale Inquiry, see: Wynne, B., *Nuclear Decision-Making—Rationality or Ritual*, British Society for History of Science, London, forthcoming 1980; Pearce, D., Edwards, L. and Beuret, G., *Decision-Making for Energy Futures: A Case Study of the Windscale Inquiry*, MacMillan, London, 1979; and Breach, I., "The Windscale Inquiry: 100 Days of History", *New Scientist*, London, 1978.

115. This legislation included the Radioactive Substances Act 1960 and the Nuclear Installations Act 1965.

116. He was assisted at the Inquiry by two assessors, Sir Edward Pochin, a former chairman of the National Radiological Protection Board, and Sir Frederick Warner, a member of the Royal Commission on Environmental Pollution.

the "burden of proof" was on the objectors; that BNFL's plans would be reviewed in detail, but it was for the objectors to prove a case against these plans, and not vice-versa.

Mr. Justice Parker raised three general questions on the first day of the Inquiry:

    a) should oxide fuel from the United Kingdom reactors be reprocessed in Britain at all;

    b) if so, should such reprocessing be carried out at Windscale; and

    c) if so, should the plant be about double the estimated size required to handle United Kingdom oxide fuels, so that the spare capacity could be used for reprocessing foreign spent fuel?

The Windscale Inquiry was in several respects a unique experience in the history of British public local inquiries. It was the first time that national issues had been explicitly included in the terms of reference of a local inquiry. It was the first time that a major, critical component of overall British energy policy had come under such detailed public scrutiny. And, it was the first time that so much scientific and technical nuclear-related information had been made generally available to the British public.

The local public inquiry format also had a number of important implications for the way the investigation was conducted. The importance Mr. Justice Parker attached to "discovery of evidence", "technical proof" and "scientific fact" reinforced the adversary structure of the Inquiry proceedings. Since the burden of proof was largely on the opposition and pressure groups and not on the proponent, BNFL, these groups required access to considerable financial resources, information and technical expertise. Moreover, despite the semi-legal nature of the proceedings, participation at the Windscale Inquiry in most instances demanded recourse to legal counsel. Finally, unlike a Planning Inquiry Commission (PIC) or a Royal Commission, the local public inquiry mechanism did not allow for extensive and in-depth independent investigation or research.[117] It was, rather, a mechanism designed for the timely review of existing information and competing arguments.

Although the Inquiry had been established in early March, it did not get underway until mid-June, 1977. During this interval, considerable confusion reigned as to its eventual scope and terms of reference. Preliminary hearings to clarify these matters were not held until mid-May, and even then it was still unclear as to exactly what issues could be raised and in what ways.[118] Several procedural issues were, however, clarified, namely: that the Inquiry would proceed in a sequential fashion, with each intervenor group having an opportunity to present testimony and cross-examine previous witnesses; that access to information would be reviewed on a case by case basis and would be made available to all parties; and that the hearings would last for approximately three months.

Nevertheless, a number of important procedural matters were not clearly resolved.[119] These related to: the substantive scope of the Inquiry; the pre-

---

117. A Planning Inquiry Commission (PIC) has never been established in Britain, although the closest approach was the one taken by the Roskill Commission into the siting of the third London airport. Some have alleged that a PIC, with its broad investigative powers for research and independent analysis, would have been a more appropriate mechanism at Windscale than the Public Local Inquiry format.

118. It should be remembered that Mr. Justice Parker's three central questions were not raised until this 17th May meeting, several weeks after BNFL and objector groups were required to submit their cases.

119. The matters raised here have the benefit of the 20/20 vision that comes with hindsight. They were not necessarily issues perceived as important by Mr. Justice Parker or the intervenor groups at the moment when Inquiry procedures were laid down.

paration of impact analyses; methods for pre-Inquiry information exchange between all the concerned parties; and the question of funding for intervenor groups.

First, in the absence of a clearly articulated and generally agreed-upon definition of the issues to be addressed at the Inquiry itself, detailed case preparation was difficult. This was as true for British Nuclear Fuels Ltd. (BNFL), as it was for the intervenor groups. In fact, BNFL had been hoping as late as December 1976, that the Inquiry would not get underway until the autumn of 1977. When it was actually announced for mid-June, BNFL was under extreme pressure to present a fully detailed application and preparation for its defence by early March. BNFL's uncertainty as to which issues it would have to defend against opponents was further accentuated by the ill-defined division of responsibilities between it and Government Department representatives as to who would be responsible for defending what aspect of Government policy at the Inquiry.

The lack of a clearly pre-defined scope for the hearing also handicapped the preparatory efforts of intervenor groups, who did not have full access to BNFL's THORP application until it was submitted on March 1, 1977. They then had only 21 days in which to indicate whether or not they would appear at the Inquiry and on what grounds they based their objections. Most groups therefore presented only the most general grounds for objection, so as to preserve the broadest possible margin of manoeuverability. However, the net result was that the substantive scope of debate was not really narrowed until the Inquiry officially opened on June 14th, and even then, focused and concise discussion was impeded by the lack of adequate preparation on both sides.

The United Kingdom does not have a formal system of environmental impact assessment as do, for example, Canada, the United States and, more recently France.[120] This means that, in the case of BNFL's application, no overall, integrated assessment of THORP-related environmental impacts was available for public discussion and comment until immediately before the Inquiry. One result was that considerable time was taken up at the Inquiry itself, clarifying and verifying information related to environmental impacts. Had such an environmental impact statement been presented for public comment and criticism many months earlier, it might not only have facilitated the Inquiry proceedings, but also served to clarify important areas of substantive disagreement for formal treatment at the Inquiry.

A further important factor was that no statutory period of information-exchange between parties was established for the critical preparatory phase of the Inquiry. This meant that both sides in the debate were at a disadvantage in preparing their cases. For the opposition groups, this was especially critical, because of the fact that, under the public local inquiry procedures, they bore the burden of proof. The result was that these groups imposed considerable information demands on BNFL, at the very moment when it was itself under extreme time pressure to complete its own case. Moreover, later during the Inquiry itself, cross-examination produced an even greater "overload" of information demands. At one stage, this pressure was so great that BNFL had over 200 individual sets of information demands outstanding.

Finally, government funding was not provided to intervenor groups before the Inquiry. This meant that each group had to secure its own source of independent financing. Since the official estimate for the Inquiry duration was

---

120. However, under the Town and Country Planning Act 1971, and the Town and Country Planning (Scotland) Act 1972, local authorities may make environmental protection a condition for land-use permits required for any development.

from eight to ten weeks, and since most groups decided to engage full-time legal counsel, whose normal fees were about $2,000 per week, the sums required were substantial. In addition, there were considerable costs associated with engaging the services of expert witnesses, some of whom came from abroad, as well as the costs of simply keeping abreast of Inquiry developments and of maintaining contact with other intervenor groups. Some of these groups such as the Network for Nuclear Concern (NNC) and the Political Ecology Research Group (PERG) were able to manage extremely cost-effective presences at the Inquiry by foregoing legal counsel in favour of a full-time unpaid advocate. Others, such as Friends of the Earth (FOE) and the Conservation Society succeeded in securing some funding by means of a newspaper campaign organised by "People for a Non-Nuclear World".

However, many groups found themselves competing with one another for funding. This competition also had a negative spill-over effect later in the Inquiry when intervenor groups were required to present expert witnesses, some of whom they had dually-engaged. Thus, funding insecurity not only limited the ability of some groups to plan ahead, but led to intergroup misunderstandings and considerable dissipation of effort at the critical moment when such groups might otherwise have been preparing their presentations for the Inquiry. As one participant subsequently noted: "This kind of uncertainty, distraction and conflict was being endured at the very time when groups should have been meeting jointly to organise a coherent case. Instead, mistrust and general confusion was more apparent".

The conduct of the Windscale Inquiry was itself hampered by the very problems that beset the intervenor groups. Some of these groups were simply not sufficiently well-prepared. They had entered the Inquiry with their cases only partially developed, hoping to be able to finalise their arguments in the light of information available from cross-examination. Many had not had the time, interest or inclination to discuss their case strategies with other intervenor groups before the Inquiry began. Others displayed a lack of internal cohesion, organisation and discipline.

These problems were further accentuated by the fact that the Inquiry itself generally proceeded by a sequence of witnesses (proponent and opponent group representatives) and not by issue. This resulted in the presentation of frequently overlapping and repetitive testimony, and even in contradictory testimony being given by intervenor groups which had not co-ordinated their strategies.

To take one example: on the issue of the relative safety of storage of spent fuel rods and reprocessing, intervenor groups found themselves embarrassingly at odds with one another in their testimony. Two groups, the Windscale Inquiry Equal Rights Committee (WIERC) and the Political Ecology Research Group (PERG) argued that because oxide spent fuel was allegedly unreliable in open storage ponds and could lead to massive aerial releases of radioactivity, spent fuel should not be imported for reprocessing. Friends of the Earth (FOE), on the otherhand, emphasized the safety of spent fuel storage, arguing that because it was safe, international fuel storage at Windscale might offer a viable alternative to reprocessing and, thus, to the risks inherent in plutonium extraction.

One of the tendencies of local public inquiries is to emphasize polarity not diversity. Their adversary format is based on the assumed existence of two diametrically opposed parties. In the case of the Windscale Inquiry, these supposedly consisted of BNFL and government departments on the one side, and the environmentalist, anti-nuclear and other associated groups on the other. However, in reality, the intervenor groups represented a broad spectrum of ideologies, motivations, strategies, styles and capabilities. They did however share a number of common concerns: about the need to reappraise the "ethic" of industrial expansion as a value in and of itself; about the dangers of technocratic domination over processes of democratic choice; about the position of THORP as a stepping-stone to a full-scale plutonium economy; and about the need for greater public accountability on the part of government and the private commercial sector as well.

Nevertheless, group perceptions as to what the Inquiry was really about and the strategies adopted for the presentation of cases differed widely. These differences explain in part the sometimes contradictory nature of opposition group testimony and the lack of a supposed "united front". For example, Friends of the Earth adopted a relatively narrow "technical" line of argument at the Inquiry, seeking, in a sense, to beat BNFL at its own game. It indicated a willingness to accept a small nuclear component as a part of the overall British energy programme, but sought to achieve a delay in the THORP decision. At the other extreme were such groups as the Windscale Appeal, SEI, SCRAM, and PERG,[121] all of which called for the outright abolition of nuclear energy per se. And, in the middle were the NNC and TCPA,[122] taking more moderate stands along the tactical lines of FOE, but at the same time sympathetic to the rationale of the other more extreme groups.

The adversary format of the Inquiry had, however, two other important consequences. It brought to the surface a considerable amount of new and potentially valuable information. Nevertheless, given the extremely rapid pace of the Inquiry, many groups were simply unable to keep fully abreast of, or adequately respond to, this information. It also reinforced the general tendency towards a curtailing of political argument in favour of adjudicating scientific and technical facts. A collateral tendency was towards reducing all possible decision options to but two: full-scale THORP development—or nothing. In effect, other options, such as delaying the decision or building a smaller THORP facility to meet just the needs of the United Kingdom, were never seriously debated before the Inquiry. Even if they had been, it is likely that considerably more time than was available would have been needed.

In the end, it was this factor—timing—that appeared to be the most crucial element. According to the managing director of BNFL, "An Inquiry of this kind should never have taken place: the Government should have identified the question of nuclear-fuel reprocessing as a major and controversial issue long before a site application had been submitted for the proposed Windscale plant".[123] Representatives of many of the opposition groups agreed: the Inquiry, it was believed, had been held too late and too hurriedly. They felt that 100 days had been a critically short period for one of the most technically diverse and complex inquiries ever held in Britain.

In the end, too, it was one man, Mr. Justice Parker, assisted by two scientific assessors, who was left to interpret the conflicting evidence, evaluate the various arguments, and reach a final decision. In early February, 1978, three months after the conclusion of the Inquiry, he presented his report to the Minister of the Environment, recommending full-scale development of the BNFL thermal oxide reprocessing plant (THORP) at Windscale.[124] He rejected unambiguously almost all the arguments raised by objectors; citing at times one group's testimony

---

121. Society for Environmental Improvement (SEI), Scottish Campaign to Resist the Atomic Menace (SCRAM), and the Political Ecology Research Group (PERG).
122. Network for Nuclear Concern (NNC) and Town and Country Planning Association (TCPA).
123. Mr. Con Allday, quoted in the *New Scientist*, 10th November, 1977, p. 332.
124. *The Windscale Inquiry (Parker Report), op. cit.*

against the other. Criticisms were raised in several quarters. As one commentary in *Nature* put it:

> "Parker was there to decide, not to illuminate controversy"... "Once (he) had decided that the decision should go in BNFL's favour he went out of his way to find for them on almost every issue—rather as a judge confronted by a bunch of witnesses prepared to testify to a man's innocence, might dismiss their evidence *in toto* once satisfied that the man was guilty".[125]

Several months later, Parliament had its opportunity of debating the findings and recommendations of Mr. Justice Parker's report. This it did with relatively little controversy and considerable dispatch, subsequently voting to approve the Government's plan to expand BNFL's reprocessing facilities at Windscale.[126] In effect, one of the undeniably significant results of the Windscale Inquiry was that it served to depoliticise and decrease public debate on Windscale.[127] For a large sector of the general public, sometimes confused by the panoply of conflicting technical facts and arguments raised at the Public Inquiry, Mr. Justice Parker's unambiguous recommendations and findings were a relief. They seemed to provide the comforting assurance that wisdom and objective scrutiny could prevail over unreasoned irrational allegations on controversial issues of extreme uncertainty and technical complexity.

For many others, however, while the Inquiry did serve to reduce uncertainty in some areas, it left many important issues still unresolved. These included the broad political issues of nuclear proliferation and energy-forecasting: two matters which were believed to be inseparably linked to any reprocessing decision, but which had yet to be the subjects of major British public policy debate. Finally, to a number of other people, the Windscale Inquiry revealed the need to develop much more effective participatory mechanisms for encouraging public debate and coping with scientific and technologically-related controversy. As the *New Scientist* commented, "nothing resembling these hearings will do for future exercises intended to resolve publicly issues of great technical depth and complexity".[128]

2. The Mackenzie Valley Pipeline Inquiry

If the Windscale Inquiry can be partially characterised as a trial between competing facts and competing logics, then the Mackenzie Valley Pipeline Inquiry was a consciousness-raising public teach-in. One was to the other like a courtroom to a school. But, only partly so. The official purpose of the Inquiry was to examine the social, economic and environmental impact of a natural gas pipeline in the Northwest Territories and the Yukon, and to recommend the terms and conditions that should be imposed if the pipeline were to be built.

Ever since the 1968 discovery of enormous oil reserves at Prudhoe Bay, Alaska, the Canadian Government had been actively promoting northern oil and gas exploration activities. Preliminary results by 1970 had indicated large reserves of natural gas in the Arctic Islands. Beginning in that year, plans were devised for a 1.22 meter diameter pipeline that would connect Prudhoe Bay

---

125. *Nature*, 23rd March, 1978, p. 297.
126. Parliamentary debates took place in late March and again in mid-May, 1978, when Parliament approved the Government's Special Development Order on Windscale by a 220 to 120 vote.
127. For example, a public demonstration, organised by Friends of the Earth in London in early May to manifest opposition to the Windscale plant, drew only 10,000 supporters. While later in the same week, over 60,000 people participated in an anti-National Front demonstration.
128. *New Scientist*, 10th November, 1977, p. 332.

in the west with the Mackenzie Delta to the east for the transportation south of natural gas to US and Canadian consumers over 5,000 kilometers away. Early estimates in 1971 indicated that the pipeline project could be well underway by 1973, and would cost less than $8 billion.

In 1972, the Canadian Government issued a set of "Expanded Guidelines for Northern Pipelines", which required pipeline applicants to submit a detailed environmental impact statement and assume the burden of proof for demonstrating that any proposed pipeline would be acceptable environmentally as well as socio-economically.[129] It also indicated that, since only one pipeline would be built, potential applicants should combine efforts. In mid-1972, a private industrial consortium was formed which came to be known as Canadian Arctic Gas Pipelines Limited, referred to as "Arctic Gas".[130] Over the next two years, Arctic Gas spent over $50 million on research and studies to support its environmental impact study. At the same time it took the unprecedented step of establishing an independent Environment Protection Board (EPB) comprised of scientists and engineers to perform an "outside" assessment of its project.[131]

During 1973, however, a number of factors intervened to postpone project planning and drastically reshape public attitudes—from optimism to pessimism and suspicion of government. The international oil embargo and subsequent quadrupling of petroleum prices forced a reappraisal of national energy supply estimates, which indicated, for the first time, a near-term potential energy supply/demand gap. This led to a lack of faith in energy forecasting, and to new concerns on the part of the public about possible national energy shortages, which were further amplified by news of the disappointing results of gas exploration activities in the Mackenzie Delta.[132] Coupled with growing public environmental awareness were revelations about alleged government mis-handling of northern offshore oil-drilling permits, which in turn prompted new concerns about the adequacy of existing environmental protection measures. Finally, and perhaps most significantly was the emergence of northern Indian and native people opposition to *any* pipeline, which raised new concern about the government's treatment of northern minority groups and their claims to land.

It was in response to these pressures and in the light of the fact that the Arctic Gas pipeline proposal was soon to be submitted for authorisation that the Government took two important initiatives in early 1974. The first was to establish an in-house government Pipeline Application Assessment Group (PAAG) to undertake a preliminary review and evaluation of the pipeline application. The second was to establish a Public Inquiry into the proposed pipeline project. It appointed Mr. Justice Thomas R. Berger, a member of the Supreme Court of British Columbia, as Commissioner. However, the Inquiry did not

---

129. Department of Indian Affairs and Northern Development, *Expanded Guidelines for Northern Pipelines*, No. 72-3, Supply and Services Canada, 28th June, 1972, pp. 25-30.
130. Among the 12 founding members were the following: Standard Oil of Ohio, Atlantic Richfield, Humble Oil and Refining (EXXON), Canadian National Railways, Pacific Lighting Gas Development, Columbia Gas Systems, Northern Natural Gas and five American and Canadian pipeline companies. Subsequent members included among others, Imperial Oil, Shell Canada and Gulf Oil Canada.
131. The EPB was provided with $3.5 million for the preparation of technical studies. Some of these study findings ultimately came to play a critical role in the subsequent Inquiry in that they indicated that several critical aspects of Arctic Gas's proposed pipeline project were environmentally unacceptable.
132. Unlike Prudhoe Bay, where huge volumes of hydrocarbons were locked in a single pool, deposits in the Mackenzie Delta appeared to be isolated in numerous smaller pockets. Of the first 66 wells drilled, four had encountered oil, 11 had discovered gas, and the remaining 5 were dry. See: Pimlott, D., *Oil Under the Ice*, Canadian Arctic Resources Committee, Ottawa, 1976, p. 12.

formally get underway for one full year so as to allow all parties sufficient time for preparation. Formal hearings began in early March, 1975, and ended 20 months later on 19th November, 1976.

One of the first things Mr. Justice Berger did was to hold a number of detailed *preliminary hearings* during April, May and September 1974, on the way the inquiry ought to be conducted.[133] These hearings highlighted a series of issues and conflicts concerning the scope and timing of the Inquiry, access to information, funding for intervenors, and the type and format of public hearings.

On the basis of testimony received at these hearings, he decided that the Inquiry should not be limited simply to the Arctic Gas proposal, but should encompass all potential activities related to the development of a Mackenzie Valley transportation corridor, and their economic, environmental, human and cultural impact. He also sought to institute a "system" of hearings that would be responsive both to the needs of the Inquiry for different types of information and to the needs of all potential participants to contribute to the best of their capabilities. Four different sets of public hearings were organised.

The first of these were *Formal Hearings*. These were established to receive evidence relating to the pipeline construction proposals, the measures to be taken to meet the guideline requirements, and related studies concerning anticipated project impacts. They were organised in four sequential issue phases: engineering and construction; physical environment; living environment; and human environment. Semi-formal rules of evidence were instituted governing presentation of evidence, calling of witnesses, cross-examination and rebuttal.

Disclosure of all studies and reports in the hands of government, industry and public interest groups (including native organisations) was made mandatory.[134] Mr. Justice Berger directed all participants to provide a list of all studies and reports in their possession or power relating to the Inquiry. These were then provided to all other participants, who were free to demand that any study or report on any list be produced.

A second set of *Special Hearings* was held to receive evidence from potential gas producers (Gulf, Shell and Imperial) as to their exploration and production programmes in the gas fields of the Mackenzie Delta and Beaufort Sea. These were aimed specifically at examining environmental risks and proposed protection measures, as well as the potential economic impact on northern communities.

The third set of hearings were the *Southern Hearings,* held in large cities across southern Canada. These were informal hearings, supposedly aimed at enabling Canadians who could not appear in the North to express their views. In effect, they achieved much more than that: they brought the issues of the North to the consciousness of people in the south to a degree that had been largely unanticipated.

The fourth type of hearings represented perhaps the most significant innovation of Mr. Justice Berger. These were the *Community Hearings*. They were held in each community in the Mackenzie Valley, the Mackenzie Delta and the Yukon likely to be affected by the construction of a pipeline and by corridor development. These informal hearings were held in school gyms, halls, tents, hunting camps and out-of-doors, in 28 cities, towns, villages, settlements and outposts across Northern Canada.

---

133. See: Berger, T.R., *Northern Frontier-Northern Homeland: Report of the Mackenzie Valley Pipeline Inquiry,* Vol. 2, Appendix 2: "Preliminary Rulings", Supply and Services Canada, Ottawa, 1977.

134. Mandatory disclosure of information is guaranteed by Section 19(f) of the *Territorial Lands Act* (RSC 1970, C.T-6) under which the Inquiry was established.

The community hearings were a significant departure from previous public inquiry procedures in several respects. In form, they were informal and unpretentious—almost to a fault. People who testified were not cross-examined. Witnesses spoke what was on their mind, often in a highly personalised way.

Mr. Justice Berger believed that the formal hearings and the community hearings should be regarded as equally important parts of the same process, and not as two separate processes. To try to achieve this, he introduced two further steps. The first, and most significant from a public information perspective, was to arrange with the Canadian Broadcasting Corporation (CBC) to provide radio summaries of the evidence given at the formal hearings to be broadcast to the Northern communities. The CBC Northern Network agreed to do this at their own expense on a regular basis, broadcasting in English and in the six native languages, thus helping to ensure that people in the communities could follow what had been said in the formal hearings and prepare to respond in subsequent community hearings.

The other step was to hold the community hearings concurrently with the formal hearings. From time to time the Inquiry would break off from one phase of the formal hearings to hold hearings in the communities. Thus, testimony from native persons was not relegated solely to the community hearing process; native witnesses were also called to testify at the formal hearings.

The second major innovation of the Berger inquiry was its financial support to intervenor groups. A number of environmental and other public interest organisations had pointed out at the preliminary hearings that they would simply not be in a position to participate effectively unless they received some form of assistance.

Mr. Justice Berger noted that the terms of reference of the Environmental Protection Board (EPB), set up by Arctic Gas, had been purposely limited to an examination of the consortium's own proposal, whereas the scope of the Inquiry was much broader. Moreover, the Government's own Pipeline Application Assessment Group (PAAG) was to be disbanded as soon as its report was finished in late-1974. He saw the need therefore for continuous, independent, across-the-board environmental expertise, and counter-expertise, at the Inquiry.

In late 1974, he met with then Minister of Indian Affairs and Northern Development, Jean Chrétien, to devise a set of criteria governing funding for intervenor groups.[135] The Government then agreed to provide $494,000 to the Berger Inquiry for support of research and studies by the environmental and other interest groups. A further $1,065,000 was also contributed directly by the Department of Indian Affairs and Northern Development to native groups.

A third innovation was that Mr. Justice Berger directed his Commission Counsel to serve as an active participant in the hearings. He empowered the Commission Counsel to obtain any evidence that he deemed relevant to the

---

135. These included the following:
1. There should be a clearly ascertainable interest that ought to be represented at the Inquiry.
2. It should be established that separate and adequate representation of that interest would make a necessary and substantial contribution to the Inquiry.
3. Those seeking funds should have an established record of concern for, and should have demonstrated their own commitment to, the interest they sought to represent.
4. It should be shown that those seeking funds did not have sufficient financial resources to enable them adequately to represent that interest, and that they would require funds to do so.
5. Those seeking funds had to have a clearly delineated proposal as to the use they intended to make of the funds, and had to be sufficiently well-organised to account for the funds.

Inquiry. The intention was again to ensure that as many facets as possible of evidence would be considered in the Inquiry. Participation by Commission Counsel, as will be seen, played a critically important role in later phases of the Inquiry.

Finally, Mr. Justice Berger sought to discourage the impression that his staff, including Commission Counsel, was a kind of "privy council" to the Inquiry, whose views and recommendations were not subject to public scrutiny. He therefore directed Commission Counsel to present publicly his arguments supporting the terms and conditions that it felt should be imposed on any pipeline right-of-way permit.

The Inquiry hearings did not begin until 3rd March, 1975. In the interim, the Government's in-house Pipeline Application Assessment Group (PAAG) released its report in late 1974, pointing out a number of weaknesses in the Arctic Gas application.[136] It also highlighted two of the central problems that were to plague the Inquiry: inadequate scientific knowledge and the lack of specific information as to how industry planned to proceed with the pipeline project.

These problems also reveal and underline a more generic difficulty related to the assessment of large-scale, technically complex projects. This was succinctly put by one observer: "In most huge technological projects, such as the Mackenzie Valley pipeline, there can be no trial runs. Problems, even failures, are demonstrable only when the entire system is in place; and the only people with the technical knowledge to evaluate the project are its proponents, who are all the more likely to err precisely because of their own vested interest and the lack of independent criticism".[137]

The fact that in late March 1975, a second group of companies, Foothills Pipelines Ltd., announced plans for an alternative gas pipeline down the Mackenzie Valley is of great significance, since the presence of two competing pipeline proposals set the stage for an adversary process in which each company was obliged to defend its proposals against the criticisms of the other. As a result, much more technical information became available during the Inquiry than would otherwise have been the case.

From the very beginning of the public hearings, it was apparent that the procedures designed to broaden public participation and promote accessibility to Inquiry findings also provided a further important dividend: they brought together two very divergent types of perceptions, values and information.

For example, one of the most important points at issue in the entire pipeline debate was the question of Indian land claims. To the industrial interests, whose pipeline projects required authorised rights-of-way across Territorial and Indian lands, this was perceived as a relatively straightforward negotiable matter: that is to say, that in exchange for such rights-of-way the native peoples would receive some monetary or other compensation. However, this perception of the matter hardly coincided with the Indians' more persuasive concept of "land". As one native spokesman explained at the community hearings:

> "Ownership does not rest in any one individual but belongs to the tribe as a whole as an entity. The land belongs not only to the people presently living, but it belongs to past generations, and the future generations that are yet to be born. Past and future generations are as much a part of the tribal

---

136. Department of Indian Affairs and Northern Development, Pipeline Application Assessment Group, *Mackenzie Valley Pipeline Assessment*, Supply and Services Canada, Ottawa, 1974.

137. Gamble, D.J., "The Berger Inquiry: An Impact Assessment Process", *Science*, Vol. 199, 3rd March, 1978, p. 948.

entity as the living generation. Not only that, but the land belongs not only to human beings but also to other living things; they too have an interest".[138]

It soon became apparent that it was the cumulative effect of development that, because of the specific and short-term incremental concerns of industry and others, was likely to be overlooked. And it was precisely this cumulative aspect that was of central concern to the native people.

Testimony at the community hearings from non-technical persons also demonstrated the fallibility of the conventional belief that only people with specialised technical knowledge should make decisions about technological matters. Time and again it was the so-called "non-experts" who provided important insights and information concerning such natural phenomena as, for example the vulnerability of the Beaufort Sea, sea-bed ice scour, and native hunting and trapping practices. This "non-expert" testimony provided the elements for a more comprehensive understanding of both quantitative and qualitative impacts of the proposed development project.

In May 1976, Mr. Justice Berger brought his Inquiry to southern Canada. The aim was to give southern Canadians a chance to express their views on the proposed pipeline. A public opinion survey undertaken just before these hearings began, indicated that 62.7 per cent of southern Canadian adults were already aware of plans to construct a pipeline through the Mackenzie Valley. It is interesting to note, that, of those asked to rank five issues in order of priority, the largest group responded that protection of the environment was their first priority concern.[139]

These southern hearings began in Vancouver (B.C.) on 10th May, and, moving eastward through eight major Canadian cities, ended in early June in Halifax, Nova Scotia. The southern news media, which up until May 1976 had reported only sporadically on the activities of the Berger Inquiry, now devoted considerable and sustained coverage to these hearings. Little new evidence concerning the possible impact of the proposed pipeline construction was presented; rather, they served primarily to draw attention to the already evident differences of opinion but, more importantly, to promote broader public awareness of the issues facing the North.

The community hearings were completed in the summer of 1976, and by early fall the Inquiry was in its final stages. Most of the environmental groups had already presented their evidence during earlier phases of the Inquiry. Of these groups, the Environmental Protection Board (EPB), which had been funded and then abandoned by Arctic Gas in early 1976, provided some of the strongest evidence contradicting industrial testimony. However, conflicting testimony from the two major industrial competitors, Arctic Gas and Foothills, brought to light a number of technical problems overlooked by these environmental groups. For example, Foothills' attack on the technical feasibility of Arctic Gas's proposed winter construction plans revealed a number of serious weaknesses in the consortium's application.

One of the most crucial technical findings resulting from the Berger Inquiry came, however, not from environmental group or industrial testimony, but from evidence presented by Commission Counsel. The Arctic Gas consortium, which had spent more than $1 million studying the problem of frost-heave at its Calgary

---

138. Mackenzie Valley Pipeline Inquiry, *Proceeding at Community Hearings*, Vol. 59, p. 6585, 26th May, 1976, Toronto, Ontario.
139. *The Globe and Mail*, Toronto, 2nd September, 1976.

test site, had testified early in 1975 that not only did they fully understand the frost-heave phenomenon and its effects on a pipeline, but that they had "complete confidence in the methods... proposed for its control".[140] These methods consisted of either burying the refrigerated pipe under three meters of earth, placing it on a built-up "berm", or both.

When Commission Counsel presented expert testimony which disputed the analysis made by Arctic Gas, the company's experts disagreed completely, pointing to their own experimental data which refuted this testimony. However, when scientists at the Canadian National Research Council tried to verify the Arctic Gas experimental results, they were unable to do so. In October 1976, Arctic Gas admitted that due to "faulty test equipment", their previous frost-heave pressure measurements were erroneous. As a result of the inquiry hearings, it became unequivocally clear that the critical problem of frost-heave—basic to the theory and design of the pipeline project—was far from being understood or resolved.

This issue not only revealed inadequacies in some aspects of the pipeline proposals, but serious gaps in the scientific and technical knowledge available for the effective assessment of any large-scale, technologically innovative undertaking. As Mr. Justice Berger concluded: "Industry proposes and the government disposes. Without such a body of knowledge, the government will not be able to make an intelligent disposition of industry's proposals now or in the future".[141]

In early November 1976, the Mackenzie Valley Pipeline Inquiry came to a close. Mr. Justice Berger presented his findings to the government in the early spring of 1977, recommending that the construction of the Mackenzie Valley Pipeline be postponed for ten years. He concluded, moreover, that, owing to potentially severe environmental damage, no pipeline should ever be built across the Northern Yukon from Prudhoe Bay in the west to the Mackenzie Valley Delta in the east. However, he noted that it was environmentally feasible to construct a pipeline from the Delta up the Mackenzie Valley to southern Canada. This delay, he thought, would be needed in order to: settle native land claims; establish new socio-economic programmes and institutions; and fill the critical gaps in information that would be required to devise adequate environmental mitigative measures, ensure fail-safe engineering designs, and cope effectively with construction problems associated with permafrost and Arctic conditions.

The *Berger Report* was published in May 1977, and became a national best-seller—something of a rarity in the annals of government Commission reports. It seemed to many people that Mr. Justice Berger had raised issues of national concern in a way and with an objectivity that was felt to be truly balanced. As one native spokesman put it, "It's the first time anybody bothered asking us how we felt".[142]

The impact of the report in political circles was equally significant. When, on 4th July, 1977, the National Energy Board presented its own report on northern gas pipelines, it recommended that an alternative pipeline route be selected along the Alaska Highway.[143] Several months later, the Canadian

---

140. Berger, T.R., *Northern Frontier—Northern Homeland*, Vol. 1, op. cit., p. 19.
141. Ibid., p. 22.
142. Ibid., p. 22.
143. National Energy Board, *Reasons for Decision: Northern Pipelines*, Vol. 1, NEB, Ottawa, June, 1977.

Parliament approved a Federal Cabinet decision to build the Alaska Highway natural gas pipeline.

These British and Canadian experiences provide a number of important insights as to the value and limitations of commissions of inquiry as one mechanism for informing policy-makers, as well as the general public, on issues of considerable scientific and technological complexity and diversity. The Mackenzie Valley Pipeline Inquiry served essentially to "politicise"—in the non-pejorative sense of the term—a conflict which had originally been framed in narrow technical and economic terms. It served to broaden the forum of national policy debate by explicitly revealing the nature of political and value choices implicit in the pipeline decision. The result of the Windscale Inquiry was in effect to depoliticise a largely emotional and political controversy by narrowing the scope of discussion to an elucidation and adjudication of competing technical facts and interpretations. It purposefully eschewed attempts to debate the merits of Government policy by limiting itself to a consideration of the facts of that policy.

These two examples reveal the degree to which personal style can influence the process of debate before a commission of inquiry. Mr. Justice Berger saw his function not solely as an arbiter in a highly technical debate, but also a kind of "listening ear" in an open public opinion forum. The openness of his style was not just a reflection of the man, but a necessary condition for full and effective native group participation.

Mr. Justice Parker also sought to achieve an entirely open and full debate of the issues before the Windscale Inquiry; and, in fact, no group was ever denied access to the forum of debate or to available information. However, Mr. Justice Parker perceived the role of his Commission in much narrower terms, that of judging scientific and technical facts, not motivations or policy intentions. He, in fact, was the final arbiter of technical disagreement and dispute in the Windscale Inquiry. As one commentator has noted:

> "The actual issues as they unfolded under cross-examination at Windscale, produced more sophistication and polarisation, leaving Parker to make up his mind upon what appeared to be arbitrary grounds. In the end, the interpretation of expert conflict (or agreement) into policy commitments will always be arbitrary, because the ultimate determinants are human evaluations in political contexts... However, there appears to be a fundamental choice between processes which acknowledge that ultimate arbitrariness openly and maturely, and cause political institutions to grasp the nettle; and, on the other hand, processes which effectively pretend that objective reality has decreed it thus, and that impartial experts—be they scientists or judges—discovered that objective reality as only they can".[144]

These two examples, from which direct comparisons cannot be drawn, nonetheless reveal some lessons for the conduct of future national inquiries. Sufficient time for the critical preparatory phases must be allowed for, and reliable preliminary information is essential if all participants in such inquiries are to have an equal opportunity of putting forward their particular arguments. Financial assistance to certain groups appears to be especially warranted to

---

144. Wynne, B., *The Windscale Inquiry* (manuscript), *op. cit.*, p. 56.

ensure a more democratic representation of views and interests before an Inquiry.[145]

Inquiries that proceed by a structural discussion of each major issue, as opposed to some more arbitrary sequence of proponent and opposition intervenor groups, not only have the advantage of reducing the presentation of repetitive testimony, but, more importantly, of developing a cumulative information base from which to assess cumulative impacts. Recourse to the adversarial approach to the examination and cross-examination of witnesses and testimony can often result in more critical information becoming available, in the more thorough and penetrating examination of competing claims, and in the clearer articulation of individual and groups' interests and biases. This is all the more true when there is but one major proponent, as was the case at Windscale. It also has its disadvantages, however, especially when it results in sometimes exclusive reliance on lawyers and technical experts to the detriment of direct citizen participation. Developing a system of more flexible formal and informal public hearings, at which citizens and experts alike can participate, represents one possible approach. More carefully structuring the treatment of issues themselves in terms of their policy-related and technically-substantive components, is another.

One of the most ubiquitous and sensitive problems of commissions of inquiry is the confusion that often exists between their policy-*implementation* and their policy-*determination* functions. In theory and supposedly in practice such commissions serve the former function. The latter is believed to be the prerogative of Parliament. And yet, as these two examples seem to indicate, such a distinction is increasingly difficult to clearly ascertain. Just as the Windscale Inquiry served to legitimise British nuclear policy, so did the Mackenzie Valley Inquiry give definite shape to a new Canadian northern development policy. Strictly speaking, of course, Parliament exercised its ultimate prerogative in both countries with the decisive power of its vote. In both cases, however, neither parliamentary body deliberated in great depth or detail on the content of those policies.

Commissions of inquiry have of course a number of limitations. They are costly and time-consuming.[146] Their effectiveness as a participatory mechanism is often heavily dependent upon the qualities of leadership and direction provided by the person or persons appointed to conduct them. Commissions of inquiry are in most cases advisory; they are not decision-making bodies per se. Their recommendations are subject to ministerial and cabinet approval. Thus, public participation in inquiry proceedings is often only indirectly, not directly, related to government decision-making. Even despite their sometimes tacit influence on policy-making, they are not completely satisfactory surrogates for the exercise of parliamentary power.

Nevertheless, within parliamentary systems of government, commissions of inquiry can complement in a number of important ways the traditional governmental machinery. They can provide a sometimes effective means for examining the implications of medium and long-term policies, for marshalling the resources of time, objectivity and expertise required for coping with issues of extreme

---

145. See below, pp. 89-90.
146. The Mackenzie Valley Pipeline Inquiry began public hearings on 3rd March, 1975 and ended on 19th November, 1976. Its final report (Vol. II) was submitted on 30th November, 1977. Total cost, including funding to intervenor groups, was $4,937,262. The Australian Ranger Uranium Environmental Inquiry, which began on 9th September, 1975 and extended into early 1977, is estimated to have cost nearly $1 million. No figures are available for the Windscale Inquiry which lasted 100 days.

scientific and technological complexity and diversity, and for providing an open forum for the expression of public opinion.

D. SPECIAL AD HOC MECHANISMS

In addition to the more institutionalised mechanisms of advisory committees, legislative hearings, and commissions of inquiry mentioned above, there is another broad set of more ad hoc approaches, which are also aimed at informing policy-makers as to public needs and wishes. One of these is the standing or special national commission.

One example of these is to be found in the United States, where, partly because of the influence of a generally more active system of Congressional standing and select committees, the formalised institution of Commissions of Inquiry does not exist. Instead, one finds a number of special study commissions, often established by Executive Order of the President to examine a wide variety of policy problems in such areas as crime, drug abuse, pornography and population growth to name several well-known examples. The purpose of such commissions has generally been to analyse available scientific data, canvas public opinion, undertake research, and provide findings and recommendations to the President and to Congress on possible approaches to alleviating or at least coping with such problems. Of more recent and different origin is the National Commission for the Protection of Human Subjects of Biological and Behavioural Research.

Established by Congress in 1974, its purpose is to gather evidence on the practice of human experimentation and to recommend policies that would balance ethical and scientific concerns.[147] The Commission is an interesting experiment in public participation in that no more than five of its eleven members can have engaged in biomedical or behavioural research. The Commission reports directly to the US Secretary of Health, Education and Welfare (HEW), and its powers are enhanced by regulations stipulating that the HEW Secretary is required to accept Commission recommendations or officially explain his reasons for rejection or modification in the US *Federal Register*. As of the Spring of 1978, the Commission had already issued six reports and sets of recommendations in such areas as foetal research, psychosurgery, research information disclosure, and research involving prisoners, children and those institutionalised as mentally ill.[148]

All the Commission's meetings are open to the public, and more than ten public hearings have been held as well as site visits to prisons and institutions for the retarded and emotionally disturbed. Two of the Commission's reports have been incorporated into the Department of HEW's regulations concerning research involving prisoners and foetus research. However, it is in this latter area of human foetuses where one witnesses some of the limits on the Commission's powers.

Public controversy over foetal research was one of the motivating reasons behind the establishment of this Commission in the first place. Beginning in late 1974, the Commission held public hearings, sponsored independent studies, and reached a near consensus on the adoption of a set of rules and research

---

147. Established under the *National Research Act* (PL 93-348) of 12th July, 1974.
148. Four additional reports appeared in 1978: Advances in biomedical and behavioural research, Basic ethical principles that should underlie the conduct of research on human subjects, The performance of Institutional Review Boards, and Application of research guidelines to the delivery of health services.

guidelines which were presented to the Secretary of HEW in May 1975. Most of these recommendations were agreed to, but the final rules were drafted independently by the Secretary and diverged on several significant points. And, in the end, it was the Secretary of HEW whose ultimate authority prevailed.

***

The emergence over the last decade of "third party" groups some of which often refer to themselves as "public interest" groups has caused no small degree of consternation and anger on the part of government bureaucrats. One of the sources of bureaucratic concern is to be found in the commonly-shared belief that: a) such so-called "public interest" groups are nothing more or less than special interest groups in disguise and b) that government, its agencies and appointed officials are both more representative of, and the most appropriate guardians for, the general "public interest". This view has stimulated a number of different experiments, based on social science research techniques, to develop more refined and systematic methods for gathering unbiased information on public preferences.

Public opinion poll surveys, in-depth interviews and other related information-gathering techniques are well-developed and practiced on a continuing basis in all countries. One technique developed and experimented with by the US Forest Service since 1972 is a system called "Codinvolve".[149] This is an applied content analysis system designed to transfer large quantities of information from written statements in diverse sources into a form that is easy to summarise for policy review. After identifying questions of significance to decision-makers, researchers code all relevant public statements appearing on a given issue in personal letters, reports, petitions and resolutions, and newspaper editorials, etc. Condensing these various statements into clusters of opinions, tables are then prepared for policy-makers which summarise the nature of public inputs in terms of the degree of support or opposition expressed for a given issue or proposed project. Secondary analysis and interpretation are possible by manual or computer-assisted techniques.

The Codinvolve system is basically conceived of as an analytical tool, which permits policy-makers to evaluate a broader spectrum of attitudes than may emerge at public hearings or in response to public announcements. Since 1972, the system has been used in more than 30 Forest Service studies to analyse well over 50,000 public inputs. Costs for Codinvolve analysis have generally run between $1 to $3 per public input processed, depending upon the complexity of issues and other factors. The technique has most frequently been employed for the preparation of environmental impact studies which, under the US National Environmental Policy Act of 1969, are required for all Federal actions significantly affecting the human environment.

The Forest Service's use of Codinvolve data on public attitudes about the use of DDT provides one example of the application and limits of this technique.[150] In its environmental impact statement prepared for the Council for Environmental Quality (CEQ), the Service requested authorisation to use DDT to protect Douglas-Firs in forests in northwestern United States against moth

---

149. See: Clark, R.N., and Stankey, G.H., "Analysing Public Input to Resource Decisions: Criteria, Principles and Case Examples of the Codinvolve System", *Natural Resources Journal* (USA), Vol. 16, No. 1, January, 1976, pp. 213-236.

150. See: Kelley and Rompa, "Public Opinion About Controlling the 1973 Douglas-Fir Tussock Moth Outbreak" (mimeo), US Forest Service, Pacific Northwest Region, Portland, 1973.

damage. It provided Codinvolve data indicating considerable public support for DDT use, owing to the general unavailability of other effective alternatives and concern about the economic implications of moth infestation. However, the CEQ and Environmental Protection Administration (EPA) ruled against using DDT, contrary to the wishes of local citizens, giving greater weight to the concerns of environmentalists, who were strongly opposed to DDT use.[151]

Such applied social science research techniques to measure public sentiment are increasingly being used to complement other more traditional means for informing decision-makers of citizen preferences. However, they are participatory only in the narrowest and most indirect sense. Letters, petitions and questionnaires are only solicited statements of beliefs; they do not necessarily represent a commitment to active citizen participation in decision-making.

Finally, one of the most active and direct manifestations of citizen preferences is, of course, the public protest. Public demonstrations and protests are, in principle, no more of an illegitimate form of public participation than are letters, petitions, resolutions and manifestos. They are, however, often more visible, extreme and sometimes violent. The problem is that they are, by their very nature, non-institutionalised and therefore often non-controllable forms of expression.

The motivations behind public demonstrations are often as diverse as are the number of public policy areas they touch. Public protests in areas related to science and technology do not appear to be that different from those encountered in other areas. At least three broad categories of reasons behind public protests can be identified. These are: to influence public opinion by politicising issues and forcing them onto the political agenda; to cause governments to reconsider the implementation of decisions or policies; and to open up the decision-making process itself to consideration of a broader representation of public views.

Public protests in any democratic society should be considered a political warning light. In many ways and within certain limits public protests can be socially beneficial, not just as a "safety valve" to let off steam, but also as a political indicator of the perceived responsiveness of government to the needs and aspirations of its people.

It is often difficult to fully assess the influence of public protests or demonstrations on any given issue or decision. However, the number of participatory reforms that governments have undertaken over the last decade, especially with respect to administrative and regulatory decision-making, is one general indication of the influence that public protests have had on government. The general thrust of these reforms has been not only to "open up" government decision-making, but to develop more effective processes for the reconciliation of conflicting special interests with those of the general public.

---

151. A year later, following a second moth outbreak, EPA authorised a one-time use of DDT.

IV

RECONCILING CONFLICTING INTERESTS

Government administrative agencies and tribunals were once commonly believed to embody the broadest interests of the general public. To many people, they still do. But along with the recent decline in public confidence in political institutions in general has crept a growing mistrust of governmental administrative and, particularly, regulatory bodies. Their role as unbiased advocates for the public interest has been undermined; their powers of objective adjudication questioned; their increasing reliance upon "technical expertise" has been seen as a derogation of responsibilities for direct public consultation.

Part of the problem resides in the fact that the "public interest" is not a unitary concept. It involves many competing interests which comprise aspects both beneficial and harmful to the "public good". The problem is further complicated by the emergence in recent years of a plethora of new minority groups representing a vast array of specialised economic, social, political, cultural and ethnic interests. In addition to these is another broad new category of "third party" groups, some of which claim to speak for the "public interest" in certain areas of public policy concern.

One of the tasks of government agencies is to provide an open and impartial forum in which these interests can be heard, to weigh the competing claims and, in striking a balance, to seek what will provide the greatest possible benefit to the largest number of people. To achieve this goal fairly, effectively and efficiently is no small task. It requires that all interests be represented in these forums where the public interest is defined and judged. Therein lies one of the most essential needs and demands for public participation.

Governments have sought to respond to these new participatory demands and needs in a number of ways. Administrative ministries and agencies have attempted to expose decision-making processes to a broader representation of views by expanding public hearing procedures and promoting more active public involvement. Regulatory boards and tribunals have begun to relax stringent requirements for "standing". In some cases public advocacy structures have been established to inform citizens when their interests are directly affected. In others, financial assistance has been provided to intervenor groups in rule-making proceedings. Recourse to administrative appeal courts, ombudsmen, and citizen litigation have become increasingly common, as citizens have sought to redress alleged abuses of administrative autonomy.

Many scientific and technologically-related public policy debates with which administrative agencies and regulatory tribunals must cope involve conflicts between technical expertise and counter-expertise. These situations are often further complicated by the difficulty inherent in trying to separate facts from values. Institutional credibility and the credibility attached to the decisions that result often depend upon the manner in which these conflicts are resolved.

One witnesses, therefore, a number of new experiments with mechanisms directed towards the adjudication of scientific and technical conflicts.

Many of these initiatives to promote broader public participation in government administrative, regulatory and adjudicatory decision-making processes are marked by a certain ambivalence. In some cases they represent little more than a gesture. This is because many government officials and politicians are hesitant about the implications of expanded direct public participation in government decision-making. They fear that it may create unnecessary delays in administrative proceedings and foment increased citizen litigation. Politicians are also concerned that it might diminish their own influence and thus weaken the system of representative democracy.

There is also the concern that participatory measures could further special interests at the expense of the general public interest, and encourage self-appointed public representatives who lack technical knowledge to assess decisions. As Brooks has noted: "I predict that the public will once again return to decision by experts if it comes to feel that the participatory processes are being used not to better define the public interest, but rather to further special interests or political ideologies out of the mainstream".[152]

A. ADMINISTRATIVE DECISION-MAKING

Laws and regulations governing administrative practices and procedures in most countries stipulate that all affected parties must be notified and consulted in the preparation of most government decisions. We have referred earlier to the different approaches that have been taken to citizen notification concerning decision-making procedures. Such efforts become all the more crucial when it is not immediately evident whose interests are, or will be, affected by a given decision, and whether in fact the effects will be of a direct or indirect nature. Public hearings provide one important means for determining the scope of concerned interests, for securing information on competing claims, and, most importantly, for trying to reconcile those conflicting demands in terms of some broad concept of the general public welfare.

Governments have undertaken a number of reforms in recent years to meet public demands for broader representation in administrative proceedings. In France, for example, public interest inquiry procedures have been recently revised to require a broader scope to inquiries, a longer duration and public disclosure of inquiry findings and conclusions.[153] The purpose of these reforms has been summarised by the Prime Minister in the following terms:

> "By giving its opinion concerning the project in this way, the public takes part in the process of assessing the "public interest" of the project submitted for inquiry, such assessment of course remaining the sole prerogative of the State in the legal context of expropriation for public purposes".[154]

---

152. Brooks, H., "Technology Assessment in Retrospect", in *Newsletter on Science, Technology and Human Values* No. 12, Harvard University, Cambridge, 1976, p. 22.

153. There are three types of public inquiry in France: "d'utilité publique", "de commodo et incommodo", and "hydraulique".

154. *Journal Officiel*, 19th May, 1976. The concept of "public interest" has itself been changed, or, as some lawyers would have it, "distorted" by recent rulings of the "Conseil d'Etat". These now refer to the concept of a "cost-benefit" balance that has become the "golden rule" of jurisdictional disputes in France with respect to expropriations. See: Affaire Epoux Ellia, 24th January, 1975, in *Chronique de jurisprudence*.

However, in France, as in several other European countries, public hearings per se, in the broad Anglo-Saxon sense of the term, do not yet exist. That is, public access is often limited to persons and groups having a "direct and substantial interest" while others with only an indirect or non-pecuniary interest may not participate in public inquiry or hearing proceedings.

However, even in the United States and Canada, where such hearings are usually open to all members of the general public irrespective of the precise nature of their interests, there is another limitation. This concerns the degree to which public hearings are responsive to all the interests of all participants. For example, while hearings provide one effective means for considering the possible impacts of administrative decisions, they are not always equally effective as a forum for discussing the merits of those decisions or the goals of governmental programmes. The experiences of the US Corps of Engineers provides one illustration of this problem.

In the early 1970s, the Corps initiated a major experiment to expand its public hearing process.[155] Because of the controversial nature of many of its projects—to supply irrigation or hydro-electric power—it had traditionally used public hearings as a means for informing the public on the nature of its proposals. However, because of public resistance over the fact that hearings were often held after plans had already been well-formulated, the Corps of Engineers devised a new "open" policy aimed at synchronising public participation with the process of project proposal development.[156]

It introduced a staggered set of three public hearings, the first held at the exploratory stage of project planning to obtain information on the scope of issues needing consideration. A second set of hearings were then called to evaluate public support for alternative options. These were followed by a third set of hearings, just prior to project implementation, to further encourage public support by allowing last minute modifications. In addition, Corps personnel organised meetings with community leaders and citizen groups in order to ensure the broadest possible representation of all interests.

These efforts to broaden public participation were lauded by many citizen groups, and even by those environmental groups which had served as the Corps' severest critics. Nevertheless, the Corps of Engineers subsequently retreated from its affirmative actions to promote participation. Although the hearings had improved the Corps' image and promoted a useful dialogue with community groups, they had not significantly reduced or prevented conflict. Despite the Corps' efforts to achieve a broader public consensus, serious disagreements persisted.

The problem, it appears, has become an almost classic one. While the Corps sought to achieve greater public acceptability for the implementation of its projects, the public sought to question the very need for the projects; the hearings provided an effective forum for discussing the former but were insufficient for coping with disagreements over the latter. As one analyst has commented:

"In the tradition of... organisational theorists, the Corps assumed that a well-intentioned, positive, open planning process would result directly in

---

155. Many of the Corps' projects come under the Federal Water Pollution Control Act which requires agencies to document their compliance with legislative mandates for public participation, and gives citizens the right to petition for public hearings if none are planned. See: Environmental Protection Agency, "Public Participation in Water Pollution Control", *Federal Register*, 23rd August, 1973, p. 163.

156. Dodge, B.H., "Achieving Public Involvement in the Corps of Engineers Water Resources Planning", *Water Spectrum Bulletin*, No. 9, 3rd June, 1973, p. 449; and Hanchey, J., *Public Involvement in the Corps of Engineers*, NTIS, Springfield, 1975.

greater satisfaction with the organisation among the participants, which would in turn foster greater congruence of goals between the individual participants and the organisation, and finally, facilitate more effective implementation of the organisation's goals. Instead, participants clearly distinguish between the open planning activities of the Corps, which they were by and large quite satisfied with, and the organisation's goals, the Corps' project proposals, which they rated relatively low". [157]

Thus, public hearings before administrative bodies often appear to be of limited utility as a participatory mechanism if essential disagreements over public needs and programme goals cannot be adequately resolved. This problem is even more acute, as we shall see, with respect to regulatory decision-making processes.

Environmental impact and assessment review procedures represent one of the most important developments in relation to public participation in many countries over the last decade. The preparation of environmental impact statements, which originated with the US National Environmental Policy Act in 1969, are now required in a growing number of countries.[158] They have been recommended by the standing Royal Commission on Environmental Pollution in the United Kingdom and are under consideration in Japan. What they require is quite simple: that any government agency proposing an action state as clearly and fully as possible what the probable impacts of that action will be and what reasonable alternatives for action are. Fulfilling this requirement is not often easy.

Each year in the United States, more than 30,000 proposed Federal actions are reviewed to determine whether or not government agencies will require an environmental impact statement (EIS). Since 1969, nearly 8,000 EIS's have been filed with the Council on Environmental Quality. Many of these statements have been criticised for their bulkiness, attention to minutiae, and lack of critical detail or insufficient treatment of alternatives.[159] Experience over the last half-decade has resulted in the development of clearer guidelines for the preparation of such statements. There has also been a trend away from individual project statements towards more generic or programmatic impact statements dealing with far-reaching and long-range decisions on such diverse subjects as the conduct of recombinant DNA research, liquefied natural gas (LNG) terminals, and the space shuttle.[160]

Such impact statements serve two important purposes: they inform decision-makers of the likely effects of proposed actions and they provide the public with the opportunity to express their views concerning those actions. They have also opened the door to considerable litigation. Between 1970 and 1975, there were 332 suits filed against Federal agencies for inadequate EIS reports, of which 64 were sustained. Despite the fact that some 95 per cent of draft

---

157. Mazmanian, D., "Participatory Democracy in a Federal Agency", in *Water Pollution and Public Involvement,* H. Doerksen and J. Pierce (eds.), 1976.

158. For a review of OECD Member country approaches to environment protection and impact statement procedures, see: *Environmental Impact Assessment,* OECD, Paris, 1979.

159. See: Bardach, E., and Pugliaresi, L., "The Environmental Impact Statement Versus the Real World", *Public Interest,* 1977, pp. 22-38.

160. Environmental Impact Statements are available to all Federal, State and Local government agencies and the general public. Following issuance of a draft EIS, the public has 45 days to comment on all aspects of the proposed project. Agencies must then respond to all comments in a final version of the statement which is then entered as evidence in administrative or regulatory proceedings.

EIS's escaped legal challenge during this period, the threat of litigation remains a powerful participatory tool.[161]

Nevertheless, for many people the contribution of environmental impact statements to more informed public debate on technologically-related policy matters outweighs their disadvantage as a tool for litigation. For example, in a country with over 50 per cent of the world's nuclear power plants, EIS-related procedures for public comment and criticism have been cited by some people as largely responsible for the degree of public acceptability of nuclear reactors in the United States today.[162]

In Canada, where both provincial and Federal governments have also adopted environmental impact statement procedures, such assessment and review programmes have been aggressively pursued as a means for broadening mechanisms for public consultation.[163] The Federal government has established, as part of its general environmental review process, a number of special environmental assessment panels to hold public hearings on issues ranging from nuclear power plant siting to liquid natural gas (LNG) transport and tidal power generation facilities.[164] One such panel recently recommended rejection of Port Granby, Ontario, as the site for a proposed uranium hexafluoride refinery and waste management facility. This recommendation was accepted by the Minister of the Environment and Cabinet.[165] Three additional sites have been proposed in Ontario (Point Hope, Blind River and Sudbury) for the refinery by its proponent, Eldorado Nuclear Ltd. Separate environmental impact statements have been completed on each of these sites and referred to the Federal Environmental Assessment Panel, which has made them available to the general public for review and comment. Public hearings were held at each of the proposed locations in late 1978.

As in the United States, Canadian environmental assessment and review procedures do not themselves have the power of prohibiting ministries from implementing any proposed action, no matter how grave the environmental consequences. These are matters that are ultimately resolved at the Federal

---

161. Council on Environmental Quality, *Environmental Impact Statements: An Analysis of Six Years Experience by 70 Federal Agencies*, GPO, Washington, D.C., March, 1976, pp. 31-32.

162. See: Jellinek, S. and Brubaker, G.L., "Public Acceptance of Nuclear Power in the United States", paper presented at the *International Conference on Nuclear Energy and its Fuel Cycle*, IAEA, Salzburg, 2nd-13th May, 1977, p. 7.

163. At the Federal level there is no legislation requiring the preparation of EIS's, but Cabinet Directives in 1973 and 1977, which established and adjusted the Federal Environmental Assessment and Review Process, require the referral of all projects having potential significant environmental impacts to the Minister of the Environment. The Federal Environmental Assessment Review Office, on behalf of the Minister, establishes independent panels to conduct formal reviews of such projects. These include the public review of Environmental Impact Statements prepared by proponents on the basis of specific guidelines issued by the panels concerned.

164. Only four Federal panel reviews have been completed to date. These are the Point Lepreau (New Brunswick) nuclear power station (1974), a Nova Scotia hydroelectric facility (1975), the Port Granby (Ontario) uranium refinery (1978), and the Shakwak Project which involved upgrading of the Alaska Highway from the Alaska border to Haines Junction in Canada (1978). An interim panel report on the proposed Alaska Highway Pipeline was issued in 1977 with a final report expected in 1979.

165. See: Environmental Assessment Panel, *Report on the Port Granby Eldorado Nuclear Ltd. Uranium Refining and Waste Management Facility*, Department of Fisheries and the Environment, Ottawa, 1978.

cabinet level and decided by vote of Parliament.[166] Similarly in France, which has recently adopted procedures for the preparation and review of environmental impact statements, final decisions on project implementation rest with the Conseil des Ministres. Citizen recourse to litigation, as a means practiced in the United States to alter or enforce decisions, is generally not pursued as much in Canada or France. Thus, to a large extent, environmental assessment and review procedures in these countries represent more a form of citizen consultation than a mechanism for direct public participation in government decision-making.

B. REGULATORY DECISION-MAKING

More formal than public hearings associated with administrative decision-making are the adjudicatory procedures required for the granting of licenses and construction certificates and for the setting of regulations and standards governing such activities as airport and pipeline construction, power plant siting and offshore oil drilling. Historically, such regulatory proceedings involved only a limited number of parties. Often there was but one proponent and a few private parties whose property or other financial interests were directly affected. Thus, for example in the case of hydro-electric power plant siting, the issues were often largely seen in terms of a bi-polar confrontation; one which required balancing public interests against private, and of adjudicating appropriate levels of compensation for private property loss or damage.

Increasingly over the last decade this bi-polar model has lost much of its meaning and relevance as an explanatory device for comprehending and coping with such forms of confrontation. Not only have regulatory boards and tribunals taken on new de facto and de jure policy roles and responsibilities but they have had to grapple with increasingly large-scale technological issues and projects whose impacts are multiple and often widespread. Growing public, political and environmental awareness coupled with greater information availability has stimulated new concerns about the nature and extent of these impacts. As new minority groups have emerged, mobilising sometimes around single issues, they have achieved access to regulatory boards and tribunals, multiplying considerably the number of complainant parties involved in such proceedings.

Such citizen groups have not always been successful in their initial interventions. Intimidated by legalistic proceedings, unable to provide adequate evidence or counter-expertise, and unprepared to cope with expert cross-examination, many groups have left such proceedings after their first engagement much more bitter and cynical than when they went in. The resulting cries of "industry capture" and "special interest control" of government regulatory boards and tribunals have been echoed from one country to the next.

In most cases such allegations have proved to be unfounded. In a few they have been justified. One of the central difficulties of many regulatory bodies has been due to their historically "cozy" and "symbiotic" relationships to the very interests they regulate. With the emergence of new intervenor groups and the growth in public appreciation of the significance and policy implications

---

166. This principle is set forth in the 1970 Speech from the Throne announcing the establishment of the Department of the Environment: "the inherent conflict of interest... between those who are seeking the exploitation of non-renewable resources and those who are charged with the responsibility of protecting the environment... are better debated and resolved by ministers in council and not by officials in any department".

of regulatory decisions, these relationships have come to be questioned. In almost all highly-industrialised countries, this has led to major regulatory body reforms. The central thrust of these reforms has been to ensure greater public accountability and broader public representation in regulation proceedings. For instance, in the United States, the Civil Aeronautics Board (CAB) has created an Office of Consumer Advocacy to represent consumer interests in rule-making. The Interstate Commerce Commission's Office of Rail Public Council represents consumer interests in the Rail Services Planning Office. It channels relevant information to affected citizens and encourages public attendance at hearings.

In an increasing number of countries, there is a trend towards requiring all regulatory hearings to be held in public. However, each country responds differently to the question of which persons or groups may participate in these hearings and what issues may be contested. In Denmark and Germany, the issue of "standing"—the right to participate as "full parties" in regulatory proceedings—is usually limited to those parties having an "individual, substantial interest". This is often defined in financial or material terms, or in terms of the person's or group's geographical proximity to the site of the proposed facility or project under regulation. This has meant that for those groups not meeting criteria for standing, the only way to become indirectly involved in regulatory disputes has been by providing assistance to those persons having a right to intervene. This tactic has been employed in Germany, and sometimes quite effectively, as a way of circumventing restrictive rules of standing in administrative appeal court proceedings.

In other countries, such as Canada, regulatory tribunals have exercised broad discretion in granting standing. The result has been that, generally speaking, standing has not so far posed a major barrier to citizen or group participation in regulatory proceedings, whilst in the United States, some of the major battles waged before regulatory tribunals have been fought over the very issue of standing. Recent legislation has helped to clarify and significantly extend citizen rights to intervene in regulatory proceedings, but most US regulatory commissions and boards still retain considerable discretionary powers to limit intervenor presentations.

One direct result of liberalising rules of standing and encouraging a broader representation of public interests is that more groups and individuals than ever before are now becoming actively involved in regulatory proceedings. This in turn has led to some concern on the part of regulatory bodies as to whom various intervenor groups represent. This concern has become especially acute with respect to "third party" and so-called "public interest" groups, who may represent some affected interests but who are not themselves directly or sometimes even indirectly affected.

Some people believe that "representativeness" is somewhat of a misnomer with respect to regulatory matters. Whom one represents is, of course, important when an individual or group purports to speak as a proxy or surrogate for others. Such knowledge of one's constituency is especially appropriate with respect to the exercise of political power in an electoral system where persons are chosen to represent the views of their fellow citizens. However, regulatory bodies are not, in principle, political in this sense: theirs is not an electoral function. Their role is to consider all views, irrespective of whether they are shared by few or by many, to decide whether each point of view has merit, and if so, to take it into account. As one Canadian public interest lawyer has explained:

"The only reason why membership information is useful—and why I generally avoid providing it in relation to the clients I represent if I can—is that a decision-maker is trying to find out how many members the group has in order to determine how much political "clout" it has, to help him to decide how much weight to give its submissions. I feel that his decision should be based more on the merits of the views being put forward than the regulator's fear of the consequences of ignoring him".[167]

As regulatory bodies have increasingly had to cope with issues of great scientific and technological diversity and complexity, their own in-house needs for technical information and expertise have grown. These needs have been especially felt in the areas of environmental impact assessment and long-range forecasting. In many countries, regulatory board mandates have been expanded to deal specifically with environmental impacts as in the case of the Norwegian Watercourse and Electricity Board (NVE).[168] In many cases, however, regulatory boards have been slow in developing the necessary staff capabilities to effectively fulfil their expanded mandates.

Regulatory boards have become increasingly engaged in the tricky business of long-range forecasting, so as to be able to establish appropriate rationales for policy recommendations. The results have not always been successful, at least in the eyes of many members of the general public. For example, in 1971 the National Energy Board of Canada predicted 300 years' reserves of domestic fossil fuel. Two years later, these forecasts were reversed to 25 years, and have more recently been further reduced to 7 to 8 years! This example is not unique; in almost all countries energy supply and demand forecasting has been the source of considerable political debate and public uncertainty.

However, the important point is that, while regulatory bodies continue to rely primarily on proponent and opposition parties to produce and develop factual materials, they are no longer confined solely to the role of evaluating these facts. They are assuming increasing responsibilities for independent fact-finding. The significance of this for public participation in regulatory proceedings is two-fold.

First, many intervenor groups become engaged in regulatory disputes not simply because of concern for environmental or other project impacts, but because they object to the very policy that justifies undertaking the project in the first place. An example of this is to be found in the Canadian natural gas pipeline controversy, where several intervenor groups before the National Energy Board (NEB) argued against the need for any pipeline at all, claiming that it should be denied in order to reinforce limitation on the rate of growth of national energy demand to 2 per cent per year. However, the NEB in its final report dismissed these arguments, stating that:

"There is as yet no commitment by the Canadian people as a whole to adapt rapidly to the life-styles that a two per cent growth rate would require. Rather, there are expressions of opinion by growing segments of the population that it would be a desirable course to follow. This is a far cry from already having federal, provincial and municipal commitments to this common goal, from having all the necessary legislation enacted and from each individual having changed his lifestyle and dispensed with

---

167. Private communication.
168. Paragraph 4a of the Norwegian Regulation Law, as amended, stipulates that in the preparation of plans for watercourse regulations due regard should be taken to "public interests being affected", including "special interests pertaining to science, culture, environmental protection and outside activities".

his former ingrained wasteful habits. In the Board's view, society does not change that quickly".[169]

The direct implication of the Board's argument is, of course, that the intervenor groups had not adequately established the factual basis necessary to support a "no pipeline now" decision. More generally, however, it reveals the degree to which technical expertise and counter-expertise have come to have a major influence, not only within the context of the traditional proponent-opponent conflict, but at the interface between intervenor groups and regulatory bodies themselves.

Second, establishing the necessary expertise to participate effectively in regulatory proceedings can be both time-consuming and costly. Many citizen groups find themselves at a disadvantage when confronting industrial proponents, who can afford to hire the best available technical expertise since the costs of their interventions are either tax-deductible or can be recovered by product price increases.[170] This has raised an important and controversial issue as to whether or not public funds should be provided to assist citizen groups in their interventions before regulatory boards and tribunals.

Both sides in this debate over funding to intervenor groups seem to agree that more knowledgeable public participation in government decision-making is desirable. However, some fear that direct government financial assistance to citizens' groups for the purposes of acquiring information and technical expertise will only lead to further citizen intervention and eventually to the disruption of governmental decision-making processes; that it will not only precipitate deadlocks, increase workloads of government agencies and courts of appeal, but add a new dimension of opposition and frustration. In sum, these people believe the public interest to be better served by government itself; by its elected and appointed officials through the exercise of their legislated and assigned responsibilities.[171]

Others do not necessarily disagree with this interpretation as to the essential function and responsibilities of government as the promoter and protector of the general public interest. They argue, however, that government must "also" always seek to ensure that identifiable special interests shall not be ignored; that while most citizens' groups cannot in fact represent the interests of the whole general public, it is in the public interest that their views and interests be put forward. Direct financial assistance to such groups is believed to be especially important in those policy areas involving complex scientific and technological issues, where individual and group interests are not often readily identifiable or fully represented.

The costs associated with citizen participation in regulatory proceedings vary widely depending upon its nature, its duration, and the complexity of the issue in question. In the United States, for example, it has been estimated that the cost of active participation in rule-making at the Food and Drug Administration (FDA) is in the range of $30,000 to $40,000. While at the US Interstate Commerce Commission (ICC) estimates have ranged as high as $100,000 for full-time participation in licensing proceedings. To be sure, in many

---

169. National Energy Board, *Reasons for Decision: Northern Pipelines,* Vol. I, *op. cit.,* pp. 2-182.
170. For example, see: Goetz, G. and Brady, G., "Environmental Policy Formation and the Tax Treatment of Citizen Interest Groups", *Law and Contemporary Problems,* Vol. 39, No. 4, 1975, pp. 211-231.
171. For representative views on each side of this debate in the United States, see: US Congress, Senate, Committee on Government Operations, *Public Participation in Government Proceedings Act of 1976, Hearings,* GPO, Washington, D.C., 1976.

of these and other similar cases, a considerable proportion of costs go for legal services and representation, owing to the fact that many regulatory bodies act as courts of record and adhere to rigorous legal procedures. Nevertheless, undertaking research, developing the necessary technical expertise and hiring experts can also be expensive. As one spokesman for the US Consumer Product Safety Commission has noted:

"In order for public participation to be meaningful, it must be of sufficient technical competence and presented in such a way that the Commission can rely on it to properly balance the input of the regulated industry... the high cost of meaningful participation in regulatory proceedings often excludes such participation".[172]

A number of US regulatory agencies have therefore sought to lower this barrier to public participation by developing compensatory programmes of financial assistance to citizen groups.[173] These include the Food and Drug Administration, the Civil Aeronautics Board, the Consumer Product Safety Commission and the Federal Energy Administration, among others. In addition, the US Toxic Substances Control Act of 1976 also authorises the administrator of the Environmental Protection Agency (EPA) to provide compensation for expert witness fees and for the costs of participating in rule-making proceedings.[174]

As one EPA spokesman has noted, referring to the contributions that citizen representatives have made to agency decision-making in general:

"They have provided facts and arguments relevant to the statutory purpose (of EPA) that would not otherwise have been urged on the agency and which the agency might not have uncovered or fully appreciated on its own".[175]

Nevertheless, there still remains considerable ambivalence about whether or not government should provide direct financial assistance to citizen groups engaged in regulatory proceedings. Legislative proposals aimed at consolidating existing US Federal compensatory programmes and authorising all agencies to reimburse citizens' groups for the costs of participating in administrative and regulatory proceedings have been recently defeated in the US Congress.[176]. In other countries such as Canada, where the government departed significantly from previous practice by funding intervenors before the Mackenzie Valley Pipeline Inquiry, the line appears thus far to have been drawn with respect to direct Federal assistance to citizens' groups in regulatory matters.[177]

---

172. Quoted in: Kennedy, E.M., "Beyond Sunshine", *Trial Magazine*, June, 1977, p. 44.

173. A number of these programmes developed in light of favourable rulings from the US Comptroller General, affirming the inherent statutory authority of agencies to implement such activities. See: *Decision of the Comptroller General*, 19th February, 1976, No. B-139703: "Costs of Intervention—Food and Drug Administration".

174. *Toxic Substances Control Act of 1976*, 15 USC, paragraph 2619, 1976.

175. *Trial Magazine, op. cit.*, p. 44.

176. See: Kennedy, E.M., *Statement in Congressional Record*, 121, 174, 20th November, 1975; and US Congress, Senate, Committee on Government Operations, *Public Participation in Government Proceedings Act of 1976, Hearings, op. cit.*

177. Since 1975, Federal and Provincial government funds have also been provided to intervenor groups before the Ontario Royal Commission on Electric Power Planning, the Lysyk Commission Inquiry into the Alaska Highway Pipeline, the Bayda Cluff Lake Inquiry into uranium mining in Saskatchewan, the Environmental Assessment Panel Inquiry at Port Granby (Ontario) and, more recently, the Ontario Royal Commission (Hartt Inquiry) on the Northern Environment.

Part of this ambivalence no doubt stems from the fears already mentioned that expanded direct public participation in regulatory decision-making may result in increased conflict, obstruction and delay. Coupled with these fears are more specific concerns, often expressed by members of private industry, that increased governmental regulation can adversely affect processes of technological innovation and discourage industrial risk-taking. As the Vice-President of Imperial Oil of Canada Limited has cautioned, "public hearings and public participation in general act as brakes, and if you apply these brakes too long, you'll discourage industrial investments".

These concerns about the possible impact of government regulations on innovation do not stop at technological development. They also extend to scientific research. As one person has commented:

"... the real damage is being done to research priorities and to the creative risk-taking of the investigators. Scientists are tending to select areas of research where money is available, such as cancer or energy research. The most productive and creative workers are stimulated to follow promising leads in directions unrelated to the subjects of their grants. The freedom to pursue these leads lies at the very core of the research process, and is the main source of its conspicuous success. Yet new and rigid rules have all but eliminated this essential flexibility and opportunity for serendipity on which original discovery depends".[178]

Concern has also been voiced in recent years about the possible need for and inherent dangers of setting limits to scientific inquiry.[179] This has led to increasing controversy over the nature of the relationships between science and society, and whether or not society should attempt to regulate the pace of scientific innovation, as well as to determine its directions.

Thus, the question of direct public participation in government regulatory activities must be seen in its broader social context. General public uncertainty as to the long-term social costs and benefits of increased government regulation is but one factor. Growing awareness of the need to reappraise the aims of science and technology policies and their relationships to the goals of other national policies is another.

C. ADMINISTRATIVE AND JUDICIAL APPEAL

Citizen recourse to the legal system as a means for resolving scientific and technologically-related conflicts is an increasingly prevalent phenomenon in some countries. This is especially evident within the area of environmental policy. Citizen groups have sought to block the construction of energy facilities, sue private industries for developing technologies in violation of environmental standards, and contest governmental enforcement of environmental regulations. Legal disputes have also developed over the issue of weather modification, at the same time that questions have been raised about the legal responsibilities of researchers working on earthquake prediction. Biomedical research and innovation has been the subject of several major legal disputes involving concerns about biohazards, the ethics of foetal research, and possible malpractice in the area of human experimentation.

---

178. Shneour, E.A., "Science: Too Much Accountability", Editorial in *Science,* Vol. 195, No. 4282, 11th March, 1977.
179. See: "Limits of Scientific Inquiry", *Daedalus,* Spring 1978, pp. 1-236.

The use of legal channels as a form of direct public participation in decision-making is stimulated by a number of motivations. For some citizen groups, legal recourse is considered the only remaining alternative to overt public protest. Dissatisfied with their influence through public hearings, inquiries and other traditional administrative and regulatory channels, they have appealed to the courts to overrule government decisions and judgments. Others have sought legal recourse as a means of clarifying constitutional ambiguities surrounding questions of personal freedom and choice, as in the application of medical technologies for the prolongation of life. Still others have sought to redress perceived inequities in the sharing of certain risks by seeking to establish legal precedents in scientific and technologically-related areas not heretofore covered by existing legislation.

For some people, the growth of judicial power has been a function of the failure of government agencies to respond to groups that have been able to mobilise considerable political resources and energy.[180] For others, it is the ineluctable result of legislative intent to increase governmental power without providing for its effective enforcement. As one US Federal Judge has remarked:

> "The pattern taking shape appears to be that of a Congress intent upon bringing Federal power to bear in an ever-widening range of human affairs, but having no better answer for the monitoring, supervision, and enforcement of that power than the employment of the Federal courts to these ends".[181]

In many countries, administrative courts and appeal boards are being called upon to adjudicate scientific and technologically-related conflicts that were previously considered solely in the province of governmental agencies and regulatory bodies. Not content to judge only the integrity of administrative processes—their fairness and adherence to statutory responsibilities—they are becoming increasingly involved in reviewing the adequacy of evidence to support administrative decisions as well as the merits of the decisions themselves.

In Germany, for example, separate administrative court decisions in 1977 resulted in the prohibition or suspension of construction activities at four proposed nuclear power plant sites in Whyl, Brokdorf, Grohnde and Mühlheim-Kärlich. In France, nuclear opposition and environmental groups have contested, before the administrative tribunal at Caen, government plans for the construction of a nuclear power plant at Flamanville. Similarly, in many other countries, and most notably in the United States, administrative courts have become engaged in adjudicating major conflicts related to the siting of oil and gas pipelines, electric generation facilities and high-tension transmission lines, airports, and other large-scale technological projects.

The increased involvement by administrative courts in reviewing governmental decisions has nevertheless met with some concern. In Germany, where the courts have used their powers of interpretation broadly, questions have been raised as to their possible interference with traditional legislative functions and responsibilities. This concern stems in part from the fact that administrative court decisions have not always been consistent. For example, in the case of the nuclear power plant at Whyl, the administrative court at Freiburg based its

---

180. See: Chayes, A., "The Role of the Judge in Public Law Litigation", *Harvard Law Review*, Vol. 89, 1976, pp. 1281-1316.
181. McGowan, "Congress and the Courts", Excerpt from address at the University of Chicago Law School, 17th April, 1975, in US Congress, Senate, Committee on Government Operations, *Public Participation on Government Proceedings Act of 1976, Hearings, op. cit.,* p. 26.

decision to prohibit construction on the grounds that adequate provision had not been made for concrete containment of the reactor, while at Würzburg in Bavaria, the administrative court considered that the safety of the proposed Grafenrheinfeld nuclear reactor was sufficiently ensured without additional concrete containment, and decided that plant construction could proceed. These apparent inconsistencies in administrative court decisions reveal the broad margin for interpretation in such matters as risk-assessment and safety. While courts may be competent to judge these matters, there is increasing feeling that such decisions are more appropriately left to elected representatives, not appointed judges.

Similarly, in the United States concern has also been expressed about court involvement in the judicial review of administrative decisions. As the chairman of the New York Continental Edison Electricity Company has noted:

> "I believe that judicial review should not extend as it now does to a review of the adequacy of the evidence to support administrative decisions... My reasons for this... are rooted in the character of the issue when economic factors are balanced against environmental factors. Such an issue is not really one upon which courts are competent to pass. It involves a choice among competing and difficult-to-measure social values. This choice is, fundamentally, more a matter of social engineering than it is of the sufficiency of the evidence. It is therefore more appropriate for decision by administrative bodies appointed by elected officials to implement legislative policies than for decision by courts".[182]

Be that as it may, the fact remains that in many people's opinion, government administrative and legislative bodies have not been able to assess adequately all the possible evidence relating to the many complex scientific and technologically-related issues, or to judge levels of social acceptability towards certain risks.

Increased judicial intervention may be one result of the failure of executive and legislative bodies to respond to certain general public needs as perceived by some specific groups. It is also one indication of the increasingly unclear separation of powers between executive and legislative and judicial bodies.[183] There is evidence that legislative bodies, in enacting far-reaching legislation, have in many cases been unable or unwilling to elaborate more than general objectives or intentions to guide the implementation of these policies. The courts, in turn, have been delegated considerable de facto discretionary authority to review and assess the adequacy of decisions on policy-implementation. In so doing, they have assumed new administrative, quasi-legislative functions. It is not therefore surprising that both legislative and administrative bodies are increasingly concerned about possible judicial infringement upon their traditional claims to power and authority. This prevalent ambiguity between judicial and administrative/legislative functions has been deliberately exploited by many citizen groups seeking more direct means of influencing government decision-making.

As in the area of regulatory matters, rules of standing have been modified and liberalised in many countries. This has permitted increasing numbers of individuals and citizens' groups to present grievances to administrative courts

---

182. Luce, C.F., "Has Environmentalism Been Worth It: Judicial Injustice", *Business and Society Review*, Winter 1977-1978, No. 24, pp. 18-19.

183. This separation of powers has never really been that clear, even in those countries where such divisions have received much veneration. What is perhaps significant today is the degree to which recent administrative court decisions, especially in the energy sector, have explicitly revealed to the public the real nature of this ambiguity.

and appeal boards.[184] In addition, environmental litigation has brought pressures to include, as cause for standing, not only damage to property rights but some of the more "intangible" health and aesthetic impacts of technological progress.[185]

In the United States, liberalised rules of standing have resulted in "inexorable pressures" to expand the circle of potential plaintiffs, giving rise to the so-called "class action" suit.[186] A class action suit has been described as

> "a reflection of our growing awareness that a host of important public and private interactions...are conducted on a routine or bureaucratised basis and can no longer be visualised as bilateral transactions between private individuals...The class action responds to the proliferation of more or less well-organised groups in our society and the tendency to perceive interests as group interests, at least in very important aspects".[187]

This class action device has permitted often large numbers of aggrieved citizens to come together to present common claims in adjudicatory proceedings. The emergence of this new form of "public law litigation" has in many ways led to a major transformation of the traditional concept of adjudication in the United States and elsewhere. It has raised far-reaching and important questions.[188] Are all class action claims representative of all group interests? How far can the group be extended? Can judgments be rendered final when, in fact, many if not most group claimants are absentees?

Class action suits and other forms of intervention in judicial proceedings raise yet another set of questions with respect to public participation. For example, if one is concerned with promoting broader direct public involvement in decision-making, how relevant is the public law litigation model to the pursuit of that aim? A first tentative response appears to be: only indirectly so. This is because citizen litigation, by its very nature, is often but an indirect form of participation: between the citizen and the judge is a legal process that often militates against direct citizen involvement. Represented by legal counsel and technical experts, by sophisticated legal briefs and technical expertise, the average citizen is often far-removed from the actual process of legal debate. In the class action suit, he is even likely to be entirely absent.

How accessible is the legal process to the average citizen, and what opportunity does it provide for him to express his concerns? The costs of litigation can have a major inhibiting influence.[189] This is especially true when

---

184. In some countries such as Denmark, more restrictive doctrines of standing still prevail. For example, in the so-called Lynetten case, concerning the construction of a large sewage disposal plant, appeals were brought by environmentalists and a local trade union on grounds that the plant would cause irreparable environmental damage. They estimated that 90 tons of heavy metals would be emitted annually in the air. However, the Environmental Board of Appeal ruled that it would not review the case because "the appeal was not lodged by persons or institutions having a right to appeal". See: Mouritsen, P.E., *Public Involvement in Denmark, op. cit.*, pp. 74-75.

185. For a review and bibliography on US citizen litigation, see: DiMento, J., "Citizen Environmental Litigation and Administrative Process", *Duke Law Journal*, Vol. 22, 1977, pp. 409-452.

186. See: Stewart, "The Reformation of American Administrative Law", *Harvard Law Review*, Vol. 88, 1975, pp. 1723-1747.

187. Chayes, A., *Harvard Law Review, op. cit.*, p. 1291.

188. *Ibid.*

189. In some countries, attorney's fees are awarded to the victor. In others, the unsuccessful party must bear all legal costs. While the so-called "American Rule", as applied in the United States, provides that parties to a law suit must bear their own fees and expenses even when they prevail on the merits of their contention. This "rule" has been further reinforced by the 1975 Supreme Court decision against exceptions to this practice. See: *Alyeska Pipeline Service Co. v. Wilderness Society* (421 US 240); and *Balancing the Scales of Justice: Financing Public Interest Law in America*, Council for Public Interest Law, Washington, D.C., 1976.

risks are high that, if unsuccessful in their appeal, the citizen or group will have to bear all legal costs. Moreover, private foundation support to public interest law firms, as found in the United States, is generally not available in most countries.[190] Groups must depend upon private donations, which are often insufficient to cover the costs of litigation and those associated with research and hiring of experts. Skilled legal representation is essential, but this often means that case preparation and its defence are predicated upon legal assessments as to the most efficacious tactics to employ, and not on public assessments of what is "important" to debate. As one US lawyer has pointed out, oftentimes, "the intervenor becomes the captive of the lawyer".

Finally, one might ask how effective is the adversary format of judicial proceedings in getting at the "truth" of scientific fact and fiction and at separating facts from values? This, of course, is one of the most essential questions for which there is no simple or clear-cut answer. Critics of the adversary method of resolving technical disputes argue that lawyers, because of their background and training, often seek to conquer "truth", not discover it; that facts are often purposefully obscured when they do not substantiate a given argument and that "truth" is sought only in the context of improving one's chances of winning a case. Others feel, to the contrary, that only by means of an adversary proceeding can the "truth" be exposed; that personal bias is a reality that shapes all interpretations of fact and that cross-examination is an appropriate means of clarifying those biases and facts. To many average citizens, such dialectical debate can have a quite numbing effect, as their simple and important fears and concerns become subsumed under the weight of legalistic argument and technical expertise.

Nevertheless, despite these important qualifications, citizen recourse to legal channels represents a significant development in the whole complex and controversial process of public participation. It is a reflection of more general tendencies to seek to reconcile important scientific and technological conflicts—with all their inherent political, socio-cultural and ethical connotations—by recasting them in a juridical frame of reference. This approach appears in some cases to have served as an effective means of resolving certain technical disputes. However, it remains to be seen whether or not such a narrow frame of reference will provide an adequate, let alone appropriate and necessary, means for exploring, illuminating and eventually reconciling such broad social conflicts.

---

190. For example, during the early 1970s the Ford, Rockefeller and Clark foundations provided "seed money" to a number of US public interest law firms on the understanding that those firms would soon be self-supporting.

V

COLLABORATIVE DECISION-MAKING

The very concept of "collaborative participation" in decision-making is imprecise. It implies both negotiation and power-sharing, as well as processes of information exchange and consultation. Although systems of representative democracy are predicated upon the delegation of decision-making power to elected elites, the actual exercice of that power is often unevenly dispersed amongst numerous elected and appointed administrative, judicial and legislative officials. One of the more explicit demands inherent in contemporary participatory movements is that these officials be held more directly accountable to the public for the decisions they make. Another is for a greater share on the part of the public in that decision-making power.

Many of the participatory procedures described above are largely directed towards making policy-makers more aware of public concerns, desires and needs. In most cases, of course, decision-making authority rests with legislative and administrative bodies which may or may not choose to act on the basis of these views. What distinguishes more collaborative modes of participation is the inclusion of representatives of the general public, not just as informants but as partners in negotiation, with some share of power to ensure that decisions taken will reflect public concerns.

Perhaps the most direct expression of such collaborative participation is the national, state or local referendum. Other less direct forms include advisory boards, commissions and public law litigation. Collaborative procedures can however also include the opportunity to evaluate the information upon which decisions are to be based and to negotiate the resolution of conflicting facts and ideas. Proposals for the creation of "Science Courts" and experiments with "Citizen Review Boards" provide examples of two quite different attempts to devise more collaborative approaches to the evaluation of policy-related scientific and technical information. Efforts have also been made, and suggestions put forward, for developing collaborative approaches to mediation in the setting of research priorities.

A. SCIENCE COURTS AND CITIZEN REVIEW BOARDS

One proposal that has received considerable attention in the United States is for the creation of a "Science Court" as an institution intended to help resolve the factual dimensions of technical disputes.[191] It would consist of a panel of

---

191. Kantrowitz, A., "Proposal for an Institution for Scientific Judgment", *Science*, 156, 12th May, 1967, pp. 763-764; and Kantrowitz, A., "Controlling Technology Democratically", *American Scientist*, 63, 1975, p. 508.

"sophisticated scientific judges" rather than the general public, whose task would be to evaluate the technical dimensions of controversial policy problems and render judgments concerning the "scientific facts". It would employ adversary procedures in which scientist "case managers" would be responsible for debating the technical issues on each side of the dispute and for cross-examining each other. The scientific judges, "unusually capable scientists having no obvious connection to the disputed issue", would then issue a judgment.

Originally proposed in the early 1960s, the Science Court concept did not receive widespread public attention until the mid-1970s, when political and scientific endorsement for the conduct of a "Science Court Experiment" was obtained,[192] after considerable critical analysis and debate.[193]

The assumption inherent in the Science Court concept is that factual claims can be debated and resolved apart from questions of "social value"; that, as one observer has put it, the factual basis of debate can be "laundered from other contaminants". Nevertheless, many people believe that trying to separate facts from values is extremely difficult, if not impossible, especially in areas of public policy debate involving science and technology. They point out that when issues are clearly factual, they are either non-controversial or can be adequately dealt with by existing non-adversary procedures.[194] They further maintain that the use of court-like procedures is no guarantee that "scientist-judges" or "experts" will be unbiased or entirely free from preconceptions.[195]

Concern has also been expressed about the authoritarian implications of Science Courts rendering final judgments on the scientific or technical content of controversial disputes.[196] Critics point out that such "judgments" could be construed as being more definitive than intended. They fear that this might give rise to expectations that conclusive technical answers can and should be found, thus reinforcing technocratic tendencies to apply largely technical solutions to essentially political problems.

Nonetheless, many of these critics feel that the adversary procedure represents one of the most attractive features of the Science Court proposal. Such a public adversary process could be made to serve as an effective means for evaluating expert opinion, identifying gaps in existing scientific and technical knowledge, and promoting a broader public dialogue.[197] As Nelkin points out:

> "The Science Court concept (is) most useful in its plan to organise a forum in which opposing parties confront each other on specific issues. But such a forum need not be restricted to scientists... the goal should not be to resolve disputes through scientific judgment, but to create a context for

---

192. Task Force of the Presidential Advisory Group on Anticipated Advances in Science and Technology, "The Science Court Experiment", *Science,* 193, 20th August, 1976, p. 653.
193. See: Casper, B.M., "Technology Policy and Democracy", *Science,* 194, 1st October, 1976, pp. 29-35; and Nelkin, D., "Thoughts on the Proposed Science Court", *Newsletter on Science, Technology and Human Values,* No. 18, Harvard University, Cambridge, January, 1977.
194. Nelkin, D., "Thoughts on the Proposed Science Court", *op. cit.,* p. 22.
195. See: Bazelon, D., "Psychiatrists and the Adversary Process", *Scientific American,* 230, June, 1974, pp. 18-23.
196. One American physicist has labeled the Science Court "an attempt to institute a Plato's Republic of Science. Not since the time of the trial of Galileo have we had a Canon court issuing pronouncements of scientific truth". See: Callen, E., "Letter to the Editor", *Science,* 193, 10th September, 1976.
197. As Casper has noted, such proceedings should be "structured so that the technical questions are isolated and discussed, but in such a way that their significance to competing holistic perspectives can be understood". See: Casper, B.M., "Technology Policy and Democracy", *op. cit.,* p. 33. See also: Casper, B. and Wellstone, P.D., "The Science Court on Trial in Minnesota", in *The Hastings Center Report* (USA), August, 1978, pp. 5-7.

discussion that will reveal the assumptions underlying different views and the multiple dimensions of policy problems that make them so difficult to resolve".[198]

It will be seen, therefore, that, as presently conceived, the Science Court per se is not an example of collaborative participation, except perhaps in the narrower sense of a collaboration between scientists and policy-makers in the definition and evaluation of the scientific and technical components of public policy issues. It represents, nevertheless, an important potential as a collaborative mechanism: one that would permit representatives of the general public, as well as those of the scientific community and of government policy-making bodies, to confront one another in an open, but structured discussion of the technical and social content of major public policy disputes. Such a discussion could contribute to the clarification of the nature of technical disagreements and their relationship to political concerns and social values. One example of such a mechanism is to be found in recent experiences with the conduct of Citizen Review Boards.

Public controversy over the conduct of recombinant DNA research has led a number of countries to establish governmental scientific advisory bodies responsible for the formulation and enforcement of research guidelines. The central aim of these guidelines is to ensure public health safety and the safety of research workers from potential biohazards associated with certain forms of recombinant DNA research. Emphasis has focused primarily upon establishing, reviewing, and monitoring methods of physical and biological containment, but not on questioning the social value of the research itself.

In the United States, where government guidelines have been established for the conduct of Federally-supported recombinant DNA research, local communities have become involved in assessing the adequacy of these measures for the protection of community health interests. City councils in Cambridge (Massachusetts), San Diego (California), Ann Arbor (Michigan) and elsewhere have taken a number of initiatives in this respect. For example, the Cambridge City Council authorised its City Manager to appoint a citizen review board to evaluate the safety procedures required by the US National Institutes of Health (NIH).

This Cambridge Experimentation Review Board was comprised of eight citizens, all of whom were non-scientists, representing a cross-section of the local community. Beginning in the fall of 1976, they met from four-to-six hours each week for four months, hearing more than 75 hours of expert testimony, visiting university laboratories, and reviewing available documentation. The review board, likening itself to a citizen's court, held a mock trial in which two teams of opposing scientists presented opening and closing remarks, and responded to questions and cross-examination from board members.

Although many members began the review with the assumption that any suspicion of risk should preclude research, they all agreed in the end that the research should continue. However, in their final report to the City Council, they proposed broader public representation on the university biohazards committees required by NIH guidelines and, in addition, recommended that a separate Cambridge biohazards committee be set up to oversee all research in the city.

In its report, the Board stressed the need for more informed public debate and participation in decision-making in research matters posing potentially-serious risks to community interests:

"Knowledge, whether for its own sake or for its potential benefits to humankind, cannot serve as a justification for introducing risks to the public

---

198. Nelkin, D., "Thoughts on the Proposed Science Court", *op. cit.,* p. 27.

unless an informed citizenry is willing to accept those risks. Decisions regarding the appropriate course between the risks and benefits of potentially dangerous scientific inquiry must not be adjudicated within the inner circles of the scientific establishment."[199]

Although Board members did not regard their review process as a prototype, but rather as a social experiment in the first stages of its development, they nevertheless considered this collaborative experiment a relative success, noting:

> "We wish to express our sincere belief that a predominantly lay citizen group can face a technical scientific matter of general and deep public concern, educate itself appropriately to the task, and reach a fair decision."[200]

The Board did, however, encounter a number of difficulties. Its review process was not efficient and considerable time was lost in trying to develop a perspective on how best to proceed. Since no operating budget had been provided by the city, planning and selection of witnesses was done on an incremental basis with no long-range view as to how to develop a cumulative base of information. Each Board member was responsible for educating himself or herself, with the result that "at least half of the membership of the citizen panel did not understand the technical debate sufficiently to pursue a line of questioning on a scientific issue".[201]

Nevertheless, the Board did establish its credibility with the community. Its recommendations received unanimous approval from the City Council and from the City Commissioner of Health and Hospitals. Moreover, as one Board member later commented, the purpose of the Citizen Review Board was not to replace or duplicate the analysis of technical experts or to infringe upon the power of the Cambridge City Council:

> "Its principal function (was) to examine and assess the locus of controversy when scientific experts disagree. It (was) a means by which scientists who question the wisdom of prevailing policies affecting the public health and safety can have recourse to another forum when they are frustrated with the internal politics of professional societies and regulatory agencies or when they feel the full dimension of the problem has not been considered."[202]

The experience of the Cambridge Citizen Review Board on recombinant DNA research provides one example of the potential and limits of collaborative forms of public participation. It contributed substantially to the resolution of a conflict which had divided the local City Council on whether or not to ban or severely restrict the type of university-based research to be carried out in the community. It also served to calm some community fears about potentially dangerous biohazards. On the other hand, the predominantly layman composition of the board and its difficulty in coming to grasps with some of the more highly technical aspects of the debate, meant that some scientific questions were not addressed in great depth.[203]

---

199. Cambridge Experimentation Review Board, "Guidelines for the Use of Recombinant DNA Molecule Technology in the City of Cambridge", *Report to the City Manager*, 5th January, 1977, p. 4.
200. *Ibid.*, p. 4.
201. Krimsky, S., "The Concept of a Citizen Court", paper presented to the Meeting of the American Association for the Advancement of Science, Washington, D.C., 13th February, 1978, p. 4.
202. *Ibid.*, p. 12.
203. Committees established by Princeton University and the University of Michigan were largely comprised of trained scientists. The reports of these committees indicate that they probed more extensively into the technical debate than did the Cambridge Citizen Review Board.

Ideally, one might have tried to achieve a more balanced mix of lay people and scientists on such a Citizen Review Board. This would not, however, necessarily have guaranteed a more probing technical discussion. In fact, it might possibly have led to scientists dominating the debate and guiding it in directions not intended or desired by lay members.

Nevertheless, recently revised Federal government guidelines for the conduct of research on DNA have been implemented in the United States to increase public participation in decisions concerning the conduct of experiments. These guidelines, which are designed to increase the pace and scope of some types of research previously believed to be unnecessarily constrained by restrictive controls, require that existing institutional "bio-safety committees" draw 20 per cent of their members from the general public having no connection with the institution where the research is being performed. In addition, membership on the existing governmental National Recombinant DNA Advisory Committee has been increased from 11 to 25, with a larger percentage of persons than in the past having no involvement in recombinant DNA research.

Similarly, in the United Kingdom, where a national Genetic Manipulation Advisory Group (GMAG) was established by the government in January 1977, attempts have been made to ensure a broad representation of interests. The Group now comprises 8 scientific and medical experts; 5 members—including its chairman—who represent the general public interest; 4 members nominated by the Trades Union Congress (TUC) to represent employee interests; and 2 members, one each nominated by the Confederation of British Industry (CBI) and by the Committee of Vice-Chancellors and Principals (CVCP), to represent the interests of management.

B. NEW MEDIATION PROCEDURES

One approach to resolving technical controversy that has received increasing attention in some countries is that of mediation. This is based upon a voluntary involvement of contesting parties who confront one another to discuss factual disagreements, often in the presence of a mediator. The role of the mediator (or mediation council) is not to impose a settlement, as in the case of arbitration, but to initiate, guide and facilitate negotiation; to ensure that potential agreements will lead to workable solutions and, in the event of no solution, to clarify the major points of disagreement.

Adapted from the labour negotiation model, such forms of mediation have apparently succeeded in reaching settlement in several environmental disputes. However, their application to conflicts over science and technology is more recent and less certain. For example, the approach inherent in the development of the Austrian nuclear energy public information campaign discussed earlier was one of modified mediation. That is to say, the Austrian Ministry of Industry in inviting technical experts from nuclear proponent and opposition groups to provide lists of key technical questions that should be answered before deciding on the national nuclear programme, was in effect serving as a mediator in the dispute. One important result of this process and of the subsequent information campaign was that many technical issues were considerably clarified. Although the campaign did not result in a clear social consensus on the national nuclear programme, it did serve to identify areas where further technical information was required and to clarify the nature of technical disputes.

Such mediation procedures usually work best when there are two major parties to a dispute—which is not always the case in many technological policy disputes, where fundamental disagreements more often stem from ideological or value positions. The fact that such positions tend to colour interpretations of factual issues is one reason why mediation is often not fully successful.

Mediation processes have also been devised for purposes other than conflict resolution. In the United States, mediation has been proposed by a number of labour unions as a means of assisting government in the identification of research priorities. For example, Leonard Woodcock, on behalf of several US labour unions, noted in testimony to the President's Biomedical Research Panel that unions and consumer groups could be helpful in "communicating to the biomedical research community areas in which research needs to be undertaken or the results of previous research applied".[204]

In the Netherlands, the government has agreed to fund, on a pilot basis, the establishment of special institutes at five university locations to mediate between university researchers and potential client groups (trade unions, environmental groups, etc.). The aim of these so-called "Science Shop" mechanisms is to promote socially-relevant Research and Development ("action research") on behalf of under-privileged groups. These institutes do not "mediate" in the same sense as referred to above, but perform the function of "research intermediaries" between researchers and their potential clients.

Under procedures adopted by the first "Science Shop" experiment at the University of Amsterdam in early 1978, community groups have been invited to submit questions and requests for technical assistance to the university science shop research committee.[205] Following review by this committee, requests are advertised in the university weekly newspaper, and staff members and students are encouraged to respond. The research committee oversees the performance of this research work and organises public discussions of results.[206]

This "Science Shop" experiment is, of course, relatively modest, depending as it does largely upon the voluntary contribution of researcher-staff time and energy. Nevertheless, it has served to encourage the growth of new communication links between university researchers and community groups, stimulate researcher awareness of community problems, and promote closer interactions between scientific and technical specialists and the general public.

C. NATIONAL, STATE AND LOCAL REFERENDA

Government recourse to referenda as a method of public consultation and, in some cases, as a means for direct citizen collaboration in decision-making is a relatively recent phenomenon as applied to areas of scientific and technological controversy. In most countries whose systems are based on representative democracy, referenda serve as a rather special "political method", usually employed only under exceptional circumstances. They are generally of two types: constitutional and legislative. Constitutional referenda are binding on decisions, whereas legislative referenda are, in principle, discretionary; final decision

---

204. See: Culliton, B., "Public Participation in Science: Still in Need of Definition", *Science*, 192, 30th April, 1976, p. 453.

205. Criteria governing the selection of proposals are that the "clients" must not be able to finance the research themselves, that they have no commercial aims and that they are able to "use the results to improve their situation or the situation of those they represent". Van Dijk and Leydesdorff, Amsterdam University's Science Shop, News Release, May, 1978.

206. Up until mid-1978, 200 requests had been received; 68 per cent coming from action groups, environmentalists and other similar organisations, 27 per cent from the trade union movement, and 4 per cent from individuals.

authority rests with parliamentary bodies. Although such legislative or consultative referenda have been used rarely in the recent history of most countries, they have become the subject of increasing attention in a few. For example, as described earlier in this study, both Austria and Denmark have taken or considered taking recourse to consultative referenda in connection with government plans for nuclear programme expansion. In addition, in the United States, state referenda on nuclear power have figured prominently in recent years. While in France, local consultative referenda and plebiscites have been used as a means to gauge levels of public acceptability to the local siting of nuclear power plants.[207]

The motivations behind government decisions to hold a referendum on a given issue are multiple and complex. Two sets of tactical motives can be identified.[208] The first of these relates to political party tactics. When an issue cuts across political party lines, threatening the basis of party cohesion or posing dangers of long-term cleavages, a party may favour a referendum as an instrument of arbitration; a means of bridging the gap between conflicting wings of a party. The second set of tactics are more issue-related. When a minority group or party calculates that its position on an issue will be defeated in a normal parliamentary vote, it may favour a referendum as a final reserve weapon; as a means of saving an otherwise lost cause.

In practice, there is often a considerable admixture of motives behind government, party and popular demand for any given referendum. Separating out prime causes or motives is often a difficult and hazardous task. This is partly due to the fact that general public attitudes toward referenda are sometimes unclear and often numerous. Depending on circumstances, some may argue, variously, that a referendum: provides a means for the people to express their mandate on an issue of vital national importance; or serves a "decision-legitimisation" function in that the will of all the people lends greater legitimacy to a decision than does the will of the elected few; or promotes greater respect for the political system and loyalty to governmental process; or, finally, serves as a lightning rod for unrest, a means for restoring calm, "rational" and reasonable discussion on issues that have become highly emotional.

Three relatively recent national experiences with referenda provide a number of insights into not only the motives behind demand for more collaborative forms of public participation, but into some of the potentials and limits of the referendum mechanism as a form of collaborative decision-making. These three experiences include the California (United States) state referendum on nuclear power held in June 1976, and the Norwegian and Danish national referenda on membership in the Common Market held in September and October 1972, respectively.[209] The purpose here is not to provide a detailed comparative analysis of these experiences, but to briefly identify several common elements and problems associated with the conduct of referenda bearing on technically complex public policy issues.

The California referendum is perhaps one of the more classic examples of public participation helping to place an issue on the political agenda. It was

---

207. For a review of national referenda in France, Italy, Switzerland and the United Kingdom, see: *Revue internationale de droit comparé*, avril-juin 1976, pp. 263-347.
208. See: Bjørklund, T., "The Referendum Campaign Concerning Norwegian Membership in the Common Market", The Central Committee for Norwegian Research, Oslo, 1978, pp. 6-7 and pp. 10-14.
209. See: Groth, A.J. and Schultz, H.G., *Voter Attitudes on the 1976 Nuclear Initiative in California*, University of California Press, Davis, 1976; Bjørklund, T., "The Referendum Campaign Concerning Norwegian Membership in the Common Market", *op. cit.*; Hansen, P., et al., *The Structure of the Debate in the Danish EC Campaign, April to October 1972*, University of Aarhus, Aarhus, 1974; and Mouritsen, P.E., *Public Involvement in Denmark*, *op. cit.*, pp. 78-97.

the direct result of citizen initiatives during 1975 and early 1976 to secure the necessary signatures, which under California law are required to place a referendum issue on the ballot. Know as "Proposition 15", this particular proposal, if passed, would have required the California electrical utilities to clearly demonstrate the complete safety of nuclear power facilities in the State.[210] Proponents argued that the proposal would ensure nuclear power plant safety, while opponents maintained that it would completely halt the badly needed nuclear supply option in California. Of all the ballot issues in this State election, more people voted on "Proposition 15" than any other issue; it was ultimately defeated by 3.98 million votes to 1.97 million (67% to 33%).

By contrast, the Norwegian referendum on entry into the Common Market was the result of a decision by the political elite to place the issue on the public agenda, and in so doing served to promote widespread public participation. The decision to hold an advisory referendum had been first mooted in late 1961 during the national election campaign and in early 1962, when the parliamentary Foreign Affairs Committee reviewed the EEC membership question for the first time. The referendum issue came up again in 1967 and still again a third time in 1970, when Parliament voted to renew Norway's application for membership in the EEC.[211] At this last time, the referendum received broad support across party-lines in the Committee. Finally, in April 1972 Parliament voted unanimously to hold a referendum, but by this time the EEC referendum campaign was informally already well underway. Thus, nearly a decade of growing political debate surrounded the decision to hold the referendum, but, as one study points out, the motives behind this decision were largely political tactical ones; "the idea of a referendum (in 1972) did not have as strong support among the voters as it did among members of Parliament".[212] In the end, however, 53 per cent of the voting public decided against EEC membership; a decision that went against the opinion of the majority of Norway's political and economic elites.

The motives behind the Danish decision to hold a referendum on the EEC issue also appear to be largely political tactical ones. Throughout the 1960s, there was generally high popular and political support for Danish entry into the EEC. However, during the final stages of negotiation with the Common Market in 1970 and 1971, opposition to entry rallied significantly. By early 1971, public opinion polls indicated only a 7 per cent margin between proponents (37%) and opponents (30%); a division that was even more clearly accentuated within the ranks of the Social Democratic party which was the first party to publicly support the referendum.[213] In May 1971, the Danish Parliament decided by a

---

210. Propositions similar or identical to the California citizens' initiative were also placed on the election ballots in six other states: Arizona, Colorado, Montana, Oregon, Ohio and Washington. All were defeated. For these election results, see: "Atomic Industrial Forum", *Nuclear Info.* (USA), December, 1976.

211. It should be remembered that the question of Norwegian and Danish membership in the Common Market was closely tied to that of Britain. When General de Gaulle vetoed British membership in 1962 and again in 1967, the question of Scandinavian accession was also placed in limbo.

212. Bjørklund, T., "The Referendum Campaign Concerning Norwegian Membership in The Common Market", *op. cit.,* p. 15. The last Gallup poll in 1962 showed a pro-membership plurality (37% for versus 31% against). However, by the time the referendum campaign officially began in June 1972, opponents of membership were in the plurality.

213. One possible motive was to try to "remove" the EEC question as an issue in the anticipated September, 1971 national elections. However, this was not achieved. The EEC issue was the subject of considerable debate. Nevertheless, the Social Democrats were still able, after the elections, to form a coalition government, but at "the expense of an unprecedented division of their parliamentary group". See: Hansen, P., et al., *The Structure of the Debate in the Danish EC Campaign, April to October 1972, op. cit.,* p. 59.

large majority to hold a referendum, the results of which would be binding on the government's decision on entry into the EEC.[214] In October, 1972, one month after the Norwegian decision against entry, the Danish electorate voted 56.7% to 32.9% to join the Common Market.

It is true, of course, that many other motives lay behind the Norwegian and Danish decisions to "ask the people" by means of a referendum on whether or not to join the Common Market. And, it is equally true that popular response to these referenda was motivated by different national factors and concerns.[215] However, one particular feature of these two experiences, in contrast to those of California, is the degree to which the original decision to hold referenda was heavily based on political party tactical motives, and not on extensive popular demands for greater public participation. These participatory demands were felt more as a result of the referenda decisions, not as the reason for them.

One argument against the use of the public referendum in areas of scientific, technical or economic complexity is that the issues themselves are too complicated; that they should be decided by competent experts or politicians, not by the public who is generally uninformed on the details of such matters. For example, Norwegian opinion poll findings in 1962 indicated that 24% of the voters were opposed to the idea of an advisory referendum, two-thirds of which expressed their opposition on the grounds that voters did not have a proper understanding on the issue and felt that members of Parliament were better suited to decide.[216] However, one also finds that such attitudes are sometimes shaped by public views towards the substantive issue in question. According to one Danish survey, 52% of those who eventually voted in favour of EEC membership expressed full agreement with the proposition that "the Common Market issue is so complicated that it is not really suited for being decided in a referendum", while 20% of this same group disagreed. By contrast, only 32% of those who voted against membership agreed with this proposition, while 41% disagreed fully that the issue was too complicated.[217] Similarly, one finds in California that the majority of those opposed to "Proposition 15" (that is, those in favour of the existing State nuclear energy programme) expressed the view that the issue was far too complicated to be decided by means of a public referendum.

It is not surprising therefore that once a decision to hold a referendum has been made, considerable emphasis is often placed upon public information activities, not just as a means for the promotion of more informed public debate, but as a tool for political persuasion. The Danish Government allocated slightly in excess of $2 million to various organisations for the development of information-related activities, such as lectures, courses, study circles and seminars. In addition, the major proponent and opponent groups organised their own fund-raising activities. It has been estimated that proponents of Danish membership in the EEC outspent their opponents by a very considerable amount.[218]

---

214. The referendum was held under the authority of Section 20 of the Constitution relating to decisions on the transfer of sovereignty to international bodies.
215. For example, in Norway highly influential agricultural and fisheries interests were strongly opposed to EEC membership, providing an important split amongst the so-called national organisation elites. Whereas, in Denmark these elites were generally more unified in support of Danish membership. As Mouritsen concludes: "Due to variations in economic structures, important interests were affected in different ways by the (EEC) issue in Denmark and Norway", op. cit., p. 97.
216. *Aftenposten,* 22nd September, 1962. Results quoted in Bjørklund, T., *op. cit.,* p. 12.
217. Petersen, N., *Folket og undenrigspolitikken,* Gyldendal, Copenhagen, 1975, p. 92.
218. Mouritsen, P.E., *Public Involvement in Denmark, op. cit.,* pp. 90-91.

In Norway, the ratio was much closer, with proponents outspending opponents 2:1. However, one important difference was the fact that the major organisation opposed to Norwegian accession to the EEC, the so-called "People's Movement Against Norwegian Membership in the Common Market" ("Folkebevegelsen mot Norsk medlemskap i Fellesmarkedet"), was able to develop a firm financial base during the crucial initial phases of the campaign.[219] The early availability of start-up capital greatly facilitated its activities, whereas the Danish EEC opposition movement appeared to have serious financial difficulties from the very outset.[220] Although the Norwegian EEC proponent movement, "Yes to EEC-action", is estimated to have spent approximately $4 million, it did not generally avail itself of volunteer support, as did the opposition "People's Movement", but often used paid help.[221] The Norwegian Parliament also appropriated approximately $250,000 to each to these main ad hoc organisations relatively late in the campaign.

One also finds in the United States that opponents to the California citizens' nuclear initiative (Proposition 15) also outspent supporters by a ratio of more than 5:1. In addition, more money was spent by both supporters and opponents to educate the public on the various sub-issues encompassed by "Proposition 15" than any other campaign topic in the June 1976 state election.[222]

It is difficult to assess the impact of the many different information activities associated with these referenda on peoples' voting behaviour. In each country, divided expertise contributed significantly to the polarisation on the issues in question. Such polarisation was especially evident in the California referendum campaign on nuclear energy. In Norway, foremost social-economists came to play a prominent role in the EEC referendum, lending certain support to the opposition "People's Movement". While in Denmark, significant polarisation was also found within the electorate. However, the fact that the largest Danish political parties and major labour, business and agricultural organisations all strongly favoured entry into the Common Market was, according to a number of analyses, the "determining factor".[223]

Although very considerable sums of money were spent in each referendum campaign for the preparation, production and dissemination of printed matter, newsletters, leaflets, advertisements and other forms of information material, it is questionable whether this form of information was that influential. Rather, study findings appear to indicate that informal and interpersonal discussion with friends, family and acquaintances served, by and large, as more influential channels of information on voting behaviour.[224]

It has been said that "votes count but resources decide"; that the decisions of powerful organisations shape society more than do the results of ordinary election campaigns.[225] Nevertheless, one aspect of referenda is that the individual

---

219. Bjørklund, T., *op. cit.,* pp. 22-23, 27.
220. Mouritsen, P.E., *op. cit.,* p. 91.
221. Bjørklund, T., *op. cit.,* p. 23.
222. Liverman, J.L., and Thorne, R.D., "Public Acceptance of Nuclear Power Generation in The United States", paper presented to *The International Conference on Nuclear Power and its Fuel Cycle,* IAEA, Salzburg, 2nd-13th May, 1977.
223. See: Mouritsen, P.E., *op. cit.,* p. 96.
224. See: Vilstrup, K., *EF Kampagnerne og deres virkning,* Federation of Danish Industries, Copenhagen, January, 1973; Bjørklund, T., *op. cit.,* p. 33; and Bjørklund, T., *Ressurser og Resultat,* Institut for Samfunnsforskning, Oslo, 1976.
225. See: Rokkan, S., "Numerical Democracy and Corporate Pluralism" in Dahl, Robert, A. (ed.), *Political Oppositions in Western Democracies,* Yale University Press, New Haven, 1966.

can sometimes be more crucial than the organisation; that the voter has the final say. But, like democracy itself, the referendum is an imperfect political method. It is subject to manipulation. It does not always serve the aims of its proponents as a means by which the public can express its will. It can be exceedingly costly as well as time-consuming. Moreover, in areas of vital national interest involving considerable scientific and technical complexity, the referendum is not always an effective means for resolving important technical disputes or uncertainty.

Public referenda can also have a conservative effect: they have often resulted in decisions to maintain the status quo and to reject proposals for reform. Many politicians and government functionaries, who sometimes view government itself as one of the more progressive elements of society, believe therefore that recourse to referenda should be considered with the utmost care. Concern is also expressed about the importance of preserving the primacy of representative systems of government, and of not ceding to fluctuating pressures for more direct forms of democratic expression.

Many of these same reservations raised about the referendum mechanism are also voiced against efforts aimed at encouraging more public participation in government decision-making. This partly explains why mechanisms for collaborative decision-making are generally limited in most countries.

VI

CONCLUSION

The purpose of this Report is to shed some light on the complex and fluctuating nature of public participatory activities, to describe different national experiences and approaches adopted to cope with this phenomenon, and to analyse the various mechanisms designed to meet new demands and needs for public participation in decisions related to science and technology.

It has, of course, a number of limitations. Most obvious is the fact that it raises many more questions than it answers. Moreover, although we have examined a number of different national experiences, we have not been able to delve deeply into the details of each one, which would be necessary for the purposes of more rigorous international comparative analysis. More important, we have also not been able to explore the precise nature of participation as seen and experienced by actual participants; we have examined the issue of public participation primarily from the perspective of government.

The very concept of "general public" is an amorphous one and the "public" that demands participation in decision-making is comprised of heterogeneous elements. The perceptions and motivations of the average citizen, concerned with the siting of power plants, industrial facilities or university research centres in his or her "own back yard" are different from those of "third party" groups, whose perspectives are more national or global and whose interests may be less directly affected. Moreover, some of these latter groups—such as labour unions, industrial forums and other institutionalised interest groups—often have access to traditional channels of representation that are denied to other "third party" groups or to individuals. These differences suggest the need for more careful analysis of the inter-relationships between the various "publics" demanding participation in decision-making.

We have also tended to focus primarily on public controversies involving large-scale technological developments. However, although the preponderant number of cases cited have dealt with energy-related matters and environmental or public health-related concerns, this was not an *a priori* choice: rather, it is a reflection of the existing nature and focus of contemporary participatory demands in our largely technology-based societies.

Demand for public participation in decision-making related to science and technology is not an isolated phenomenon. It is but one manifestation of the growing desire on the part of many people for more immediate and direct involvement in the conduct of public affairs. In local communities and industrial firms, in universities and other institutions, citizens are more willing today to make their views heard and to attempt to influence decisions.

A larger proportion of the population than in the past now can afford time to participate in public affairs; they are better educated and more politically aware than their forebears. They perceive the passage of events in the world around them with a heightened sense of "actualité", as a result of present-day systems of rapid communications and instantaneous media coverage. They no longer rely exclusively on their electoral vote or political party involvement to ensure that their interests or concerns are taken care of, but seek more direct means of influencing decisions and policies by affiliation to organisations and interest groups outside the framework of parliamentary institutions.

Many people perceive traditional channels of representation as having become oligarchical. There is a widespread feeling that political parties no longer reflect contemporary socio-economic conditions, and are ill-adapted to respond to emerging social needs, interests and values. People are questioning their faith in the persons elected or appointed to represent and protect their interests. They are expressing concern about trends towards increased concentration of economic and political power; power which is perceived by some as resting too much in the hands of too few. People are therefore seeking new ways of expressing their discontent, of imposing their views, and of reasserting their individual and collective influence on matters which they perceive as affecting their lives.

Nevertheless, the concept of public participation itself remains unclear. This conceptual vagueness serves many purposes. It can help to mask disaccord over substance with apparent agreement over form. It can serve the best political intentions without endangering political alliances. It can make token and manipulative gestures appear as genuine commitments to negotiation. It can also serve to stimulate demands for fundamental changes in the ways decisions are made.

Public participation is not, however, a universal phenomenon. There is, at the same time, a greater degree of "activism" on the part of some groups and individuals, and increasing apathy among others. Neither does this participatory phenomenon touch all areas of public policy: it is fluctuating, sometimes issue-oriented and often apparently random.

A. THE NATURE OF PARTICIPATORY PHENOMENA

Scientific and technologically-related issues share a number of characteristics that distinguish them from other areas of controversy. Perception of the disparities and risks which can accompany technological change and developments in science is becoming more and more explicit, and the need for greater participation on the part of the public in decision-making on such matters increasingly felt.

Science and technology are ubiquitous in all areas of life in industrialised countries: they are at the forefront of change, but are accompanied by unanticipated and undesired consequences. Some of their impacts are irreversible and their dimensions and duration unknown. Because of their novelty and complexity, the general public does not fully understand such developments or their potential implications. Their scale, intricacy and interdependence are in some cases of an order of magnitude never before encountered. Moreover, because some scientific and technological issues affect all people, the ethical and value dilemmas they raise are often of transcendent social importance and hence of growing controversy.

Traditional approaches to governmental decision-making on scientific and technologically-related matters are another cause of growing public concern and demand for citizen participation. This concern is however related to the very nature of overall governmental decision-making. Declining public confidence in many social institutions has resulted in the questioning of governmental authority and legitimacy. The growth of bureaucratic power and the apparent decline of parliamentary influence on governmental decision-making have led to a demand for a greater degree of public accountability. Limited public access to some decision-making forums and to information in the hands of government have raised questions concerning the basis upon which governmental decisions are made.

Because some issues fall within the province of several governmental departments, people perceive a bureaucratic tendency to "parcel out" parts of each issue for special treatment by separate governmental departments. One result is that it is difficult to arrive at an overall perspective with respect to the nature and cumulative impact of many scientific and technologically-related decisions. Another is a feeling that there is an absence of adequate mechanisms or public forums in which citizens could confront policy-makers in open discussion concerning the goals of governmental policy and their manifold implications.

The general thrust, then, of participatory demand would appear to be for a greater degree of public accountability; freer public access to technical information; more timely consultation on policy options; a more holistic approach to the assessment of impacts: all of which amounts, of course, to more direct public participation in the exercise of decision-making power.

Moreover, many people feel that the tendency within government departments to define broad political problems in narrow technical terms often means that those lacking the necessary "expertise" tend to be excluded from technical discussions. Many believe that public access to technical information is usually only granted to those with a demonstrable "special interest" or "competency" on a given issue, while it is denied to others whose interests are of a more general or "political" nature. While some feel that heavy government dependency upon technical expertise, as a supposedly "objective" path to truth, results in political choices being submerged or pre-empted.

For some advocates of public participation, the essential challenge lies in establishing the weight of political considerations in the processes of decision-making in matters related to science and technology. They do not suggest a diminuation in the use of scientific and technical expertise nor, of course, a rejection of factual analysis in arriving at scientific and technologically-related decisions. Rather, they contend that a better balance must be struck in considering facts and values, in assessing technical feasibility and political acceptability, in judging the validity of specific claims.

B. INFORMATION AND PUBLIC UNDERSTANDING

Government response to public demand for greater participation in decision-making has primarily focused on providing increased access to information. Several countries have begun to liberalise laws granting freedom of access to information in the hands of government. Some have introduced new measures designed to inform the public on opportunities for involvement in decision-making; or have devised information programmes and campaigns aimed at improving levels of public knowledge in areas of scientific and technological complexity. The assumption behind these efforts has been that increased information would lead to improved understanding and to more informed public debate.

The fact that this has not always been the case is perhaps not surprising. Despite extensive government efforts to "enlighten" public opinion, such measures have often resulted in increased public uncertainty, confusion and conflict. There appear to be several reasons for this, which may provide some insights for the design and implementation of future efforts aimed at better informing the general public on matters related to science and technology.

One important distinction that needs to be made is between information consciously and purposefully provided by government and that which is withheld, but to which the public has, in fact, legitimate access. In the majority of cases studied, it is this latter type of information that has most preoccupied citizen groups and individuals. In some instances, involving for example controversies over oil, gas and nuclear energy policies, the question of how these issues have been defined by government has had important consequences for the type and form of information made available.

As we have seen from the different national experiences with government energy information campaigns and committees, public perceptions of the most important issues for debate and discussion do not always coincide with those of government. One result has been that a considerable amount of technical information has been made available which has scarcely been of direct interest and relevance to the average citizen. As Thurber's little girl commented, "This book tells me more about elephants than I ever wanted to know".

Timing is a crucial factor when it comes to informing the public on scientific and technologically-related matters. The need to inform the public as early as possible on the content and implications of scientific and technological matters is clear. But of course this is not easy or simple. In many areas of technological development, design problems are not resolved until later in the developmental process. Similarly, in many areas of scientific research the lack of scientific knowledge makes it difficult to assess completely all potential impacts. The earlier the public is involved in such discussions the greater are the demands placed on its ability and willingnes to accept scientific and technical uncertainties. This is not to suggest that the public be involved and informed only after "all the facts are in". Rather, it points to the need for more continuous involvement, not just after issues have become politicised and opinions polarised, but at the stage when policy goals and objectives are being formulated.

Governments have made recourse to public information campaigns and committees for a variety of purposes: to open up and broaden public debate, to defuse controversy, postpone decisions, to lend legitimacy to past and pending decisions, and to try to achieve a broader political consensus. They have sought to interest, educate, inform and persuade the general public. This admixture of intentions has sometimes led to increased confusion and suspicion; and has resulted in some frustration and diminished effectiveness. Educating the general public on complex scientific and technical matters is no easy task. Nor is government necessarily the most effective instructor. The difficulties encountered in this respect suggest the need to identify more clearly the purpose and most appropriate mechanisms for promoting more informed public opinion.

Providing for the widest possible distribution of information on scientific and technological issues, whether from government or the scientific community, is essential. However, as we have seen, this is a necessary but insufficient condition for more informed public debate. The question of public access to information is just as crucial, and sometimes even more problematic with respect to the private sector. Industrial concern for the protection of proprietary information and trade secrets is sometimes justified, especially in technological sectors where competition is keen. However, important issues concerning public access to industrial information are raised. These relate especially to cases where only one industrial proponent is involved, and where scientific and technological information is not already available to government.

Technical information and expertise are considered to be a requisite for effective public participation. This is especially evident in the examples we have looked at in connection with citizen interventions before governmental administrative and regulatory tribunals. Citizen access to technical expertise varies widely both within and between countries. These differences are especially reflected in the attitudes expressed within the scientific community in some countries. For example, one finds in the United States a very active involvement of some scientists in many areas of public controversy. The development of a "public interest" science movement has meant that many environmentalist and other citizen activist groups have had access to scientific and technical counter-expertise not available to their counterparts in other countries. The availability of alternative or counter-expertise is important not only for citizen "opposition" groups, but for policy-makers and, especially, parliamentarians. Without it, politicians can become captive of the government bureaucracy and its selected expertise. Providing for counter-expertise can contribute significantly to the maintenance of democratic processes and serve as a check against what some people see as technocratic tendencies in government.

National experiences with promoting more informed public debate suggest therefore above all the need for more pluralistic sources of information on science and technology. Existing governmental, scientific and industrial organisations and affiliate institutions have not been able to respond adequately to growing public demands and needs for such information. The media have increasingly important functions and responsibilities to perform in this area, as do educational institutions at all levels.

However, as we have attempted to show in this Report, information and the promotion of a more informed citizenry is but one facet of public participation. Equally important is the possibility and opportunity for citizens to express themselves in decision-making processes and forums. Public demand for such forms of participation pose a new challenge to representative government.

C. PARTICIPATION IN DECISION-MAKING

Paul Valery's maxim that "all politics is based on the indifference of most of those concerned, without which politics would be impossible", is being tested in many countries today as increasing numbers of citizens are demanding a voice in policy decisions. However, this challenge cannot be measured simply in terms of numbers but in the growth of the competing demands placed before government policy-makers. One result of the so-called "dilution of the general political public", assumed to be the basis of democratic institutions in the past, is the emergence of a fragmented public, a public comprised of many individualistic and conflicting interests. These competing pressures are being especially felt today in areas associated with scientific and technological controversy.

Government response to participatory demand for a direct say in decision-making has been cautious. Some countries have liberalised rules granting greater public access to decision-making bodies, some have experimented with new public consultation techniques and have introduced new or expanded public hearing procedures. For the most part, however, these initiatives have been limited and generally passive. Moreover, in very few countries do formal participatory rights exist; most decisions on who can participate in decision-making are left to the discretion of individual governmental administrative, regulatory and legislative bodies.

Government agencies have generally reacted to participatory demands, not anticipated them. The need for new approaches and more flexible institutions for public participation in the development and formulation of policies related to science and technology appears to be especially crucial. This is to be seen not only in the area of energy policy, where fundamental questions are being raised about the inter-relationships between values associated with energy consumption and conservation, but in areas of emerging public policy concern related, for example, to informatics, biomedicine and sociobiology.

Public demand for more direct participation in government regulatory decision-making, in particular, is felt in many countries in such matters as licensing, standard-setting and assessing the adequacy of government safety measures. Citizen litigation and public referenda are two of the most direct and "effective" forms of public participation in decision-making related to science and technology. However, as we have seen, they do not always provide the most satisfactory means for resolving complex scientific and technological controversies. For many people, attempts to resolve such deep social conflicts according to the principle of "winner takes all" is far too simplistic an approach. And, for many politicians and government officials, such forms of direct participation are sometimes considered as threatening to the concept and "effective" functioning of representative systems of democracy.

This partly explains why governments have generally been hesitant about providing direct financial assistance to citizens' groups for the purposes of developing their own base of scientific or technical knowledge and expertise. And yet, as we have seen in the case of public inquiries, in particular, and in some national science information programmes, such financial assistance can serve to redress certain imbalances and inequities among participating social groups and interests. This indicates that increased and careful attention should be given to ascertaining the types of conditions and criteria under which government financial assistance might be provided.

All this suggests that a number of questions need to be examined in further detail. For example, what might be the long-term effects of increased public participation on scientific and technological innovation and on the development and orientation of research policies? What impacts will it have on public support for basic research? Can one, and indeed to what degree should one, try to achieve "closure" in public debates in areas of scientific controversy where neither available knowledge nor prevailing and emerging social values provide an adequate basis for action? And, how can one seek to achieve broader social consensus on public acceptability to risks whose full dimensions and impacts may never be known?

This Report has focused primarily upon describing various governmental mechanisms and approaches to the promotion of more informed public debate. Many of the initiatives taken to date are very tentative and are still in early stages of evolution. It is therefore too soon to evaluate fully their efficacy or impacts on governmental structures and processes. Further detailed national and comparative studies on the nature and impact of public participatory activities on government, as well as on the scientific community and on industry, could provide insights of considerable value.

Finally, however, it may be that the most important and long-term implications may lie elsewhere; on psychological attitudes and on public perceptions, feelings and emotions. Why some people participate while others abstain, why some political systems can adapt and others cannot, why some issues cause controversy and others do not—these are questions that cannot be avoided if one is seeking to understand the essential nature of public participation and its implications for government and for society-at-large.

Annex I

LIST OF PARTICIPANTS
AT MEETINGS ON PUBLIC PARTICIPATION IN DECISION-MAKING RELATED TO SCIENCE AND TECHNOLOGY

16th-17th December, 1976
16th-17th January, 1978

Mr. Rolf Berger
President, Technical University of Berlin (Germany)

Mr. Stuart Blume[1]
University of London (United Kingdom)

Mr. Hans P. Clausen
Aarhus University (Denmark)

Mr. John Crecine[1]
Carnegie-Mellon University (USA)

Ms. Beate von Devivere[1]
Battelle, Frankfurt (Germany)

Mr. Francis Fagnani[2]
Institut national de la santé et de la recherche médicale (France)

Mr. Robert F. Keith
University of Waterloo (Canada)

Ms. Dorothy Nelkin
Cornell University (USA)

Mr. Ernest Petric[2]
University of Ljubljana (Yugoslavia)

Mr. Michael Pollak
Cornell University (USA) (former member OECD Secretariat)

Mr. Pierre Thuillier[1]
Université de Paris VII (France)

Mr. Thomas Christian Wyller
University of Oslo (Norway)

Mr. Brian Wynne[2]
University of Lancaster (United Kingdom)

OBSERVERS

Mr. John Bell[2]
Science Attaché, Australian Delegation to OECD

Mr. André Oudiz[2]
Ingénieur, Commissariat à l'Energie Atomique (France)

---

1. Attended only the meeting on 16th-17th December, 1976.
2. Attended only the meeting on 16th-17th January, 1978.

Annex II

REFERENCES

The following list of references is meant to serve as a guide to available information on the subject. Most of it relates to material published since 1970.

BOOKS

Bachrach P., *The Theory of Democratic Elitism: A Critique*, Little Brown, Boston, 1967.

Belford T., *Stacking the Deck*, Common Cause, Washington, D.C., 1976.

Benello G. and Roussopoulos D., *A Case for Participatory Democracy: Some Prospects for the Radical Society*, Viking Press, New York, 1972.

Birnbaum P., Lively J. and Parry G., *Democracy, Consensus and Social Contact*, Sage Publications Ltd., London, April 1978.

Capitant R., *Démocratie et participation politique dans les institutions françaises, de 1875 à nos jours*, Bordas, Paris, 1972.

Carnot E., Nelson and Pollack D.K., *Communication Among Scientists and Engineers*, Health Lexington Books, Lexington, 1970.

Council for Public Interest Law, *Balancing the Scales of Justice: Financing Public Interest Law in America*, Washington, D.C., 1976.

Crozier M., Huntington S.P. and Watanuki J., *The Crisis of Democracy*, New York University Press, New York, 1975.

Ebbin J. and Kasper R., *Citizen Groups and the Nuclear Power Controversy: Uses of Scientific and Technological Information*, MIT Press, Cambridge, 1974.

Gordon D.D. and Mayer N.L., *Organisational Democracy, Participation and Self Management*, Lexington, 1976.

Greer-Wootten B. and Mitson L., *Nuclear Power and the Canadian Public*, Institute for Behavioural Research, York University, Toronto, 1976.

Groombridge B., *Television and the People*, Penguin Education Special, London, 1972.

Groth A.J. and Schultz H.G., *Voter Attitudes on the 1976 Nuclear Initiative in California*, University of California Press, Davis, 1976.

Groupement de scientifiques pour l'information sur l'énergie nucléaire, *Electro-nucléaire : Danger*, Le Seuil, Paris, 1977.

Habermas J., *Toward a Rational Society*, Heinemann, London, 1971.

Hanchey J., *Public Involvement in the Corps of Engineers*, NTIS, Springfield, 1975.

Hansen P., et al., *The Structure of the Debate in the Danish EC Campaign—April to October, 1972: A Study of the Opinion-Policy Relationship*, Institute of Political Science, Aarhus, 1974.

Holden B., *The Nature of Democracy*, Nelson, London, 1974.

Ippolito D., Walker T. and Kolson K.L., *Public Opinion and Responsible Democracy*, Prentice Hall, Englewood Cliffs, 1976.

Kramer D., *Participatory Democracy: Developing Ideals of the Political Left*, Sorenkman, Cambridge, 1972.

Lenoir Y., *Technocratie française*, J.J. Pauvert, Paris, 1977.

Litt E., *Democracy's Ordeal in America: A Guide to Political Theory and Action*, Dryden Press, Hinsdale, 1973.

Lucas J.R., *Democracy and Participation*, Penguin, Harmondsworth, 1976.

Nelkin D. (ed.), *Controversy: Politics of Technical Decisions*, Sage Publications Ltd., London, 1979.
Nelkin D., *Jetport: The Boston Airport Controversy*, Rutgers University, Transaction Books, New Brunswick, 1975.
Nelkin D., *Technological Decisions and Democracy: European Experiments in Public Participation*, Sage Publications Ltd., London, 1977.
Niga J., *La démocratie directe*, La pensée universelle, Paris, 1973.
Pateman C., *Participation and Democratic Theory*, University Press, Cambridge, 1970.
Plamenatz S.P., *Democracy and Illusion: An Examination of Certain Aspects of Modern Democratic Theory*, Longman, London, 1973.
Prewitt K. and Stone A., *The Raling Elites: Elite Theory, Power and American Democracy*, Harper and Row, New York, 1973.
Puiseux L., *La Babel nucléaire*, Editions Galilée, Paris, 1977.
Purcell E., *The Crisis of Democratic Theory: Scientific Naturalism and the Problem of Value*, University of Kentucky Press, Lexington, 1973.
Radford, *Bureaucracy and Participation*, Rand McNally, Chicago, 1969.
Skoie Hans (edit.), *Scientific Expertise and the Public*, Conference sponsored by the International Council for Science Policy Studies and the Institute for Studies in Research and Higher Education, Oslo, 1979 (Vol. 5).
Vanek, J., *The Participatory Economy: An Evolutionary Hypothesis and a Strategy for Development*, Cornell University Press, Ithaca, 1971.
Verba J. and Nie N., *Participation in America: Political Democracy and Social Equality*, Harper and Row, New York, 1972.
Watson, R.A., *Promise and Performance of American Democracy*, Wiley, New York, 1972.
Wheeler M., *Lies, Damn Lies and Statistics: The Manipulation of Public Opinion in America*, Norton, New York, 1976.
Wright J.D., *The Dissent of the Governed: Alienation and Democracy in America*, Academic Press, New York, 1976.

REPORTS

Attorney General's Department, Interdepartmental Committee, *Policy Proposals for Freedom of Information Legislation*, AGPS, Canberra, November 1976.
Berger T.R., *Northern Frontier—Northern Homeland: Report of the Mackenzie Valley Pipeline Inquiry*, Vols. 1 and 2, Supply and Services Canada, Ottawa, 1977-1978.
Berrefjord O., *Big Industry and Environmental Protection*, The Central Committee for Norwegian Research, Oslo, 1977.
Bjørklund T., *The Referendum Campaign Concerning Norwegian Membership in the Common Market: A Case Study on Public Involvement*, The Central Committee for Norwegian Research, Oslo, 1977.
Boasberg, Hewes, Finkelstein and Klores, *Implications of NSF Assistance to Nonprofit Citizen Organizations. Report to the National Science Foundation*, NSF, Washington, 1977.
British Association for the Advancement of Science, *Science and the Media*, BAAS, London, 1976.
Bundesministerium Forschung und Technologie (BMFT), *Kernenergie und Ihre Alternativen*, BMFT, Bonn, 1977.
Business International, *Industrial Democracy in Europe*, Geneva, 1974.
Cambridge Experimentation Review Board, *Guidelines for Use of Recombinant DNA Molecule Technology in the City of Cambridge, Report*, Office of the City Manager, Cambridge, 1976.
Council on Environmental Quality, *Environmental Impact Statements: An Analysis of Six Years Experience by 70 Federal Agencies*, GPO, Washington, D.C., March 1976.
Department of Indian Affairs and Northern Development, *Expanded Guidelines for Northern Pipelines No. 72-3*, Supply and Services Canada, Ottawa, 28 June, 1972.
Dubas O. and Martel L., *Science, Mass Media and the Public*, 3 volumes, Ministry of State for Science and Technology, Ottawa, 1973, 1975, 1977.
Federal Environmental Assessment Panel, *Report on the Port Granby Eldorado Nuclear Ltd. Uranium Refining and Waste Management Facility*, Department of Fisheries and the Environment, Ottawa, 1978.

Fox R.W., *Ranger Uranium Environmental Inquiry, Report I and II,* AGPS, Canberra, 1976 and 1977.

Garnåsjordet P.A. and Haagensen K., *Public Involvement in Hydro-Electric Power Plant Planning,* The Central Committee for Norwegian Research, Oslo, 1977.

International Atomic Energy Agency, *International Conference on Nuclear Power and its Fuel Cycle,* Salzburg, IAEA, 1977.

Jackson R.W., *Human Goals and Science Policy: Background Study No. 38,* Science Council of Canada, Ottawa, 1976.

Law Reform Commission of Canada, *Commissions of Inquiry,* Administrative Law Working Paper 17, LRCC, Ottawa, 1977.

Ministry of Housing and Local Government (Department of the Environment), *People and Planning,* HMSO, London, 1969.

Mouritsen P.E., *Public Involvement in Denmark,* University of Aarhus and the Danish Research Administration, Aarhus, 1977.

Parker J., *The Windscale Inquiry Report,* HMSO, London, 1978.

Pearce D., et al., *Windscale Assessment and Review Project—Draft Interim Report,* Social Science Research Council, London, 1978.

Porter A., *A Race Against Time: Interim Report on Nuclear Power in Ontario,* Royal Commission on Electric Power Planning, Toronto, 1978.

Radcliffe Committee, *Report of the Committee of Privy Councellors on Ministerial Memoirs,* HMSO, London, 1975.

Roberts J., *Legislation on Public Access to Government Documents (Government Green Paper),* Secretary of State, Ottawa, June 1977.

Royal Commission on Australian Government Administration, *Report and Appendices,* Vol. 1-4, AGPS, Canberra, 1976 and 1977.

Royal Commission on Environmental Pollution, *Nuclear Power and the Environment: 6th Report,* HMSO, London, 1977.

US Congress, House of Representatives, Committee on Science and Technology, *Authorizing Appropriations to the National Science Foundation: Report No. 95-98,* GPO, Washington, 18th March, 1977.

US Congress, Senate, Committee on Government Operations, *Public Participation in Government Proceedings Act of 1976: Hearings,* GPO, Washington, 1976.

Université des sciences sociales de Grenoble, Institut de recherche économique et de planification (IREP), *Le débat nucléaire en France,* CORDES/CNRS, Paris, mai 1977.

ARTICLES

Anderson S., "Public Access to Government Files in Sweden", *American Journal of Comparative Law,* XXL, 3, Summer 1973.

Apter E.D., "Modèles économiques et participation", *Futuribles,* été 1976.

Arnstein S.R., "A Ladder of Citizen Participation", *Journal of the American Institute of Planners,* Vol. XXXV, No. 4, July 1969.

Bardach E. and Pugliaresi L., "The Environmental Impact Statement Versus the Real World", *The Public Interest,* No. 49, Fall 1977.

Bazelon D., "Psychiatrists and the Adversary Process", *Scientific American,* 230, June 1974.

Binder L., "Political Participation and Political Development", *American Journal of Sociology,* Vol. 83, No. 3, November 1977.

Boeker E., "Public Information on Science and Technology: the Dutch Case", *Science and Public Policy,* December 1977.

Boffey P. and Wade N., "The Nuclear Debate, Clashes in Congress and California", *Science,* 191, 9th January, 1976.

Boss J.F. et Kapferer J.N., "Le public et la vulgarisation scientifique et technique: Enquête sur les attitudes et les comportements des Français", *Le Progrès scientifique,* 190, septembre-octobre 1977.

Carrol J., "Participatory Technology", *Science,* 19th February, 1971.

Carruthers J. "A Hidden Aspect of the Coming Nuclear Debate", *Science Forum*, December 1977.

Casper B.M., "Technology Policy and Democracy", *Science*, 194, 1st October, 1976.

Castles F., "Political Function of Groups in Sweden", *Political Studies*, XXL, 1st March, 1973.

Castles F., "Swedish Social Democracy", *Political Quarterly*, 46, No. 2, April 1975.

Chayes A., "The Role of the Judge in Public Law Litigation", *Harvard Law Review*, May 1976.

Clark I., "Expert Advice in the Controversy about Supersonic Transport in the United States", *Minerva*, October 1974.

Clark R.N. and Stankey G.H., "Analysing Public Input to Resource Decisions: Criteria, Principles and Case Examples of the Codinvolve System", *Natural Resources Journal*, Vol. 16, No. 1, January 1976.

Coates J., "Public Participation in Technology Assessment", *Technology Assessment Activities of the National Science Foundation*, GPO, Washington, D.C., 1974.

Culliton B., "NIH to Open Budget Sessions to Public", *Science*, 192, 9th April, 1976.

Culliton B., "NSF Trying to Cope with Congressional Pressure for Public Participation", *Science*, 191, 23rd January, 1976.

Culliton B., "Public Participation in Science: Still in Need of Definition", *Science*, 192, 30th April, 1976.

Daadler H. and Irvin, "Interests and Institutions in the Netherlands", *Annals of the American Academy of Political and Social Science*, No. 413, May 1974.

Daetwyler J.J., "L'information scientifique dans les quotidiens suisses", *24 Heures*, Lausanne, 13 juin 1976.

Derian J.L., "L'énergie nucléaire et l'exercice de la démocratie", *Futuribles*, N° 3, été 1975.

Dienel P.C., "Pour de nouvelles structures de participation", *Futuribles*, N° 7, été 1976.

DiMento J., "Citizen Environmental Litigation and Administrative Process", *Duke Law Journal*, Vol. 22, 1977.

Doderlein J., "Nuclear Power, Public Interest and the Professional", *Nature*, Vol. 264, 18th November, 1976.

Dodge B.H., "Achieving Public Involvement in the Corps of Engineers Water Resources Planning", *Water Spectrum Bulletin*, No. 9, 3rd June, 1973.

Dorfer N.H., "Science and Technology in Sweden: The Fabians Versus Europe", *Research Policy*, Vol. 3, No. 2, 1974.

Drago R., "Etude de droit comparé sur la pratique référendaire: France, Royaume-Uni, Suisse, Italie", *Revue internationale de droit comparé*, avril-juin 1976.

Elvander M., "Interest Groups in Sweden", *Annals of the American Academy of Political and Social Science*, 413, May 1974.

Fagence M.T., "The Design and Use of Questionnaires for Participation Tactics in Town Planning: Lessons from the US and UK", *Policy Science*, 5 (3), September 1974.

Ferniot J., "Sur-information et sous-information", *Revue des travaux de l'Académie des sciences sociales et politiques*, premier trimestre 1975.

Gamble D.J., "The Berger Inquiry: An Impact Assessment Process", *Science*, Vol. 199, 3rd March, 1978.

Garrigues C., "De la communication de masse à la communication de groupe: Expériences pour une prospective de la communication sociale", *Futuribles*, No. 16, juillet-août 1978.

Goetz G. and Brady G., "Environmental Policy Formation and the Tax Treatment of Citizen Interest Groups", *Law and Contemporary Problems*, Vol. 39, No. 4, 1975.

Grima A.P. and Wilson-Hodges, "Regulation of Great Lakes Water Levels: the Public Speaks Out", *Journal of Great Lakes Research*, Vol. 3, No. 3-4, December 1977.

Hansen S.B., "Participation, Political Structure and Concurrence", *The American Political Science Review*, 69 (4), December 1975.

Heberlein T.A., "Some Observations on Alternative Mechanisms for Public Involvement: the Hearing, Public Opinion Poll, the Workshop and the Quasi-Experiment", *Natural Resources Journal*, Vol. 16, No. 1, January 1976.

Nelkin D., "The Political Impact of Technical Expertise", *Social Studies of Science*, 5, 1st January, 1975.

Nelkin D. and Pollak M., "The Politics of Participation and Nuclear Debate in Sweden, the Netherlands and Austria", *Public Policy*, 25, No. 3, Summer 1977.

Hirsch H. and Nowotny H., "Information and Opposition in Austrian Nuclear Energy Policy", *Minerva*, Vol. XV, Autumn-Winter, 1977.

Hohenemser C., Kasperson R. and Kates R., "The Distrust of Nuclear Power", *Science*, 196, 1st April, 1977.

Holdren J., "The Nuclear Controversy and the Limitations of Decision-Making by Experts", *Bulletin of Atomic Scientists*, 32, March 1976.

Holton G. and Morison R.S. (eds.), "Limits of Scientific Inquiry", *Daedalus*, Spring 1978.

Horowitz D.L., "The Courts as Guardians of the Public Interest", *Public Administration Review*, No. 2, March-April 1977.

Huntington S.P., "The Democratic Distemper", *Public Interest*, 41, August 1975.

Ionescu G., "Responsible Government and Responsible Citizens", *Political Studies*, 23, June-September 1975.

Junck R., "La communication sociale: Pour de nouvelles structures démocratiques", *Futuribles*, No. 7, été 1976.

Kantrowitz A., "Controlling Technology Democratically", *American Scientist*, 63, 1975.

Kantrowitz A., "Proposal for an Institution for Scientific Judgment", *Science*, 156, 12th May, 1967.

Kauffmann K. and Shorett A., "A Perspective on Public Involvement in Water Management Decision-Making", *Public Administration Review*, Vol. 37, No. 5, September-October, 1977.

Kempf H. et Toinet M.F., "Le débat nucléaire français", *Etudes*, janvier 1976.

Kennedy E.M., "Beyond Sunshine", *Trial Magazine*, June 1977.

Kernaghan K., "Responsible Public Bureaucracy: A Rationale and a Framework for Analysis", *Canadian Public Administration*, Vol. 16, 1973.

King J., "A Science for the People", *New Scientist*, 16th June, 1977.

Kloman E.K. (ed.), "A Mini-Symposium: Public Participation in Technology Assessment", *Public Administration Review*, Vol. 35, No. 1, January-February 1975.

Kolata G., "Freedom of Information Act: Problems at the FDA", *Science*, 189, 4th July, 1975.

Lagadel P., "L'étude d'impact: Instrument d'évaluation des décisions lourdes", *Futuribles*, No. 9, 1977.

Lambright H., "Scientists and Government: A Case of Professional Ambivalence", *Public Administration Review*, No. 2, March-April 1978.

La Porte T. and Metlay O., "Public Attitudes Towards Present and Future Technologies: Satisfactions and Apprehensions", *Social Studies of Science*, 5, November, 1975.

Laurent P., "Mieux comprendre les réalités nucléaires", *Projet*, No. 121, juin 1978.

Lloyd C. and Irland, "Citizen Participation: A Tool for Conflict Management on the Public Land", *Public Administration*, May-June 1976.

Lucas A.R., "Legal Foundations for Public Participation in Environmental Decision-Making", *Natural Resources Journal*, Vol. 16, No. 1, January 1976.

Macrae D., "Science and the Formation of Policy in a Democracy", *Minerva*, April 1973.

Marx J., "Science and the Press: Communicating with the Public", *Science*, 9th July, 1976.

Mazmanian D., "Participatory Democracy in a Federal Agency", Doerksen H. and Pierce J. (eds.), *Water Pollution and Public Involvement*, 1976.

Mazur A., "Disputes Between Experts", *Minerva*, 11th April, 1973.

Milch J.E., "Feasible and Prudent Alternatives: Airport Development in the Age of Public Protest", *Public Policy*, Winter 1976.

Miller D., "Democracy and Social Justice", *British Journal of Political Science*, Vol. 8, January 1978.

Nelkin D., "Ecologists and the Public Interest", *Hastings Centre Report*, Vol. 6, No. 1, February 1976.

Hewitt G., "The Effect of Political Democracy and Social Democracy on Equality in Industrial Societies: A Cross-National Comparison", *American Sociological Review*, Vol. 42, 1977.

Hextra G., "Environmental Education and Public Awareness", *Planning and Development in the Netherlands*, 1973.

Nisbet R., "Public Opinion Versus Popular Opinion", *Public Interest*, 41, August 1975.

Olsen Baden, "Legitimacy of Social Protest Actions in the United States and Sweden", *Journal of Political and Military Sociology*, August 1974.

Onibokun A.G. and Curry M., "An Ideology of Citizen Participation: the Metropolitan Seattle Transit", *Public Administration Review*, Vol. 36, No. 3, May-June 1976.

Paglin M.D. and Shor E., "Regulatory Agency Responses to the Development of Public Participation", *Public Administration Review*, No. 2, March-April 1977.

Passow S., "Stockholm Planners Discover People Power", *Journal of American Institute of Planners*, 39, 1st January, 1973.

Rohr J., "Initiative populaire et décision politique", *Annales de la Faculté de droit et de science politique*, 11, 1974.

Rosen T.L., "Office of Citizen Response: The Denver Experience", *Public Administration Review*, Vol. 37, No. 5, September-October 1977.

Ruin D., "Participatory Democracy and Corporatism: the Case of Sweden", *Scandinavian Political Studies*, 1974.

Schroeder D., "The Nuclear Debate Moves to Rural Areas", *Science Forum*, December 1977.

Sigelman L. and Vanderbok W.G., "Legislators, Bureaucrats and Canadian Democracy: The Long and the Short of It", *Canadian Journal of Political Science*, September 1977.

Smith P. and Ruud S., "The Nuclear Energy Debate in the Netherlands", *Bulletin of Atomic Scientists*, 32, February 1976.

Surrey J., "Opposition to Nuclear Power: A Review of International Experience", *Energy Policy*, Vol. 4, No. 4, December 1976.

Stewart R.B., "The Reformation of American Administrative Law", *Harvard Law Review*, Vol. 88, 1975.

Taviss I., "Survey of Popular Attitudes Towards Technology", *Technology and Culture*, 13th October, 1972.

Toffler A., "La démocratie prospective", *Futuribles*, No. 7, été 1976.

Tribe L.H., "Technology Assessment and the Fourth Discontinuity—The Limits of Instrumental Rationality", *Southern California Law Review*, Vol. 46, June 1973.

Umpleby S.A., "Is Greater Citizen Participation in Planning Possible and Desirable?", *Technological Forecasting and Social Change*, 4, 1972.

Van Meter E., "Citizen Participation in the Policy Management Process", *Public Administration Review*, December 1975.

Wade N., "Freedom of Information Officials Thwart Public Rights to Know", *Science*, 175, 4th February, 1972.

Williams R. and Bates D.V., "Technical Decisions and Public Accountability", *Canadian Public Administration*, Vol. 20, No. 1, Spring 1977.

Wynne H.R.E., "Why Nuclear Option is Not Yet Ready for Full-Scale Adoption by Canada", *Science Forum*, No. 9, August 1976.

OECD SALES AGENTS
DÉPOSITAIRES DES PUBLICATIONS DE L'OCDE

ARGENTINA – ARGENTINE
Carlos Hirsch S.R.L., Florida 165, 4° Piso (Galería Guemes)
1333 BUENOS-AIRES, Tel. 33-1787-2391 Y 30-7122

AUSTRALIA – AUSTRALIE
Australia & New Zealand Book Company Pty Ltd.,
23 Cross Street, (P.O.B. 459)
BROOKVALE NSW 2100   Tel. 938-2244

AUSTRIA – AUTRICHE
Gerold and Co., Graben 31, WIEN 1. Tel. 52.22.35

BELGIUM – BELGIQUE
LCLS
44 rue Otlet, B1070 BRUXELLES. Tel. 02-521 28 13

BRAZIL – BRÉSIL
Mestre Jou S.A., Rua Guaipà 518,
Caixa Postal 24090, 05089 SAO PAULO 10. Tel. 261-1920
Rua Senador Dantas 19 s/205-6, RIO DE JANEIRO GB.
Tel. 232-07. 32

CANADA
Renouf Publishing Company Limited,
2182 St. Catherine Street West,
MONTREAL, Quebec H3H 1M7 Tel. (514) 937-3519

DENMARK – DANEMARK
Munksgaards Boghandel,
Nørregade 6, 1165 KØBENHAVN K. Tel. (01) 12 85 70

FINLAND – FINLANDE
Akateeminen Kirjakauppa
Keskuskatu 1, 00100 HELSINKI 10. Tel. 65-11-22

FRANCE
Bureau des Publications de l'OCDE,
2 rue André-Pascal, 75775 PARIS CEDEX 16. Tel. (1) 524.81.67
Principal correspondant :
13602 AIX-EN-PROVENCE : Librairie de l'Université.
Tel. 26.18.08

GERMANY – ALLEMAGNE
OECD Publications and Information Centre
4 Simrockstrasse
5300 BONN   Tel. 21 60 46

GREECE – GRÈCE
Librairie Kauffmann, 28 rue du Stade,
ATHÈNES 132. Tel. 322.21.60

HONG-KONG
Government Information Services,
Sales and Publications Office, Beaconsfield House, 1st floor,
Queen's Road, Central.   Tel. 5-233191

ICELAND – ISLANDE
Snaebjörn Jónsson and Co., h.f.,
Hafnarstraeti 4 and 9, P.O.B. 1131, REYKJAVIK.
Tel. 13133/14281/11936

INDIA – INDE
Oxford Book and Stationery Co.:
NEW DELHI, Scindia House. Tel. 45896
CALCUTTA, 17 Park Street. Tel. 240832

ITALY – ITALIE
Libreria Commissionaria Sansoni:
Via Lamarmora 45, 50121 FIRENZE. Tel. 579751
Via Bartolini 29, 20155 MILANO. Tel. 365083
Sub-depositari:
Editrice e Libreria Herder,
Piazza Montecitorio 120, 00 186 ROMA. Tel. 674628
Libreria Hoepli, Via Hoepli 5, 20121 MILANO. Tel. 865446
Libreria Lattes, Via Garibaldi 3, 10122 TORINO. Tel. 519274
La diffusione delle edizioni OCSE è inoltre assicurata dalle migliori
librerie nelle città più importanti.

JAPAN – JAPON
OECD Publications and Information Center
Akasaka Park Building, 2-3-4 Akasaka, Minato-ku,
TOKYO 107. Tel. 586-2016

KOREA - CORÉE
Pan Korea Book Corporation,
P.O.Box n° 101 Kwangwhamun, SÉOUL. Tel. 72-7369

LEBANON – LIBAN
Documenta Scientifica/Redico,
Edison Building, Bliss Street, P.O.Box 5641, BEIRUT.
Tel. 354429–344425

MALAYSIA – MALAISIE
University of Malaya Co-operative Bookshop Ltd.
P.O. Box 1127, Jalan Pantai Baru
KUALA LUMPUR   Tel. 51425, 54058, 54361

THE NETHERLANDS – PAYS-BAS
Staatsuitgeverij
Verzendboekhandel
Chr. Plantijnstraat
'S-GRAVENHAGE   Tel. nr. 070-789911
Voor bestellingen: Tel. 070-789208

NEW ZEALAND – NOUVELLE-ZÉLANDE
The Publications Manager,
Government Printing Office,
WELLINGTON: Mulgrave Street (Private Bag),
World Trade Centre, Cubacade, Cuba Street,
Rutherford House, Lambton Quay, Tel. 737-320
AUCKLAND: Rutland Street (P.O.Box 5344), Tel. 32.919
CHRISTCHURCH: 130 Oxford Tce (Private Bag), Tel. 50.331
HAMILTON: Barton Street (P.O.Box 857), Tel. 80.103
DUNEDIN: T & G Building, Princes Street (P.O.Box 1104),
Tel. 78.294

NORWAY – NORVÈGE
J.G. TANUM A/S
P.O. Box 1177 Sentrum
Karl Johansgate 43
OSLO 1   Tel (02) 80 12 60

PAKISTAN
Mirza Book Agency, 65 Shahrah Quaid-E-Azam, LAHORE 3.
Tel. 66839

PORTUGAL
Livraria Portugal, Rua do Carmo 70-74,
1117 LISBOA CODEX.
Tel. 360582/3

SPAIN – ESPAGNE
Mundi-Prensa Libros, S.A.
Castelló 37, Apartado 1223, MADRID-1. Tel. 275.46.55
Libreria Bastinos, Pelayo, 52, BARCELONA 1. Tel. 222.06.00

SWEDEN – SUÈDE
AB CE Fritzes Kungl Hovbokhandel,
Box 16 356, S 103 27 STH, Regeringsgatan 12,
DS STOCKHOLM. Tel. 08/23 89 00

SWITZERLAND – SUISSE
Librairie Payot, 6 rue Grenus, 1211 GENÈVE 11. Tel. 022-31.89.50

TAIWAN – FORMOSE
National Book Company,
84-5 Sing Sung Rd., Sec. 3, TAIPEI 107. Tel. 321.0698

THAILAND – THAILANDE
Suksit Siam Co., Ltd.
1715 Rama IV Rd.
Samyan, Bangkok 5
Tel. 2511630

UNITED KINGDOM – ROYAUME-UNI
H.M. Stationery Office, P.O.B. 569,
LONDON SE1 9 NH. Tel. 01-928-6977, Ext. 410 or
49 High Holborn, LONDON WC1V 6 HB (personal callers)
Branches at: EDINBURGH, BIRMINGHAM, BRISTOL,
MANCHESTER, CARDIFF, BELFAST.

UNITED STATES OF AMERICA – ÉTATS-UNIS
OECD Publications and Information Center, Suite 1207,
1750 Pennsylvania Ave., N.W. WASHINGTON, D.C. 20006.
Tel. (202)724-1857

VENEZUELA
Libreria del Este, Avda. F. Miranda 52, Edificio Galipán,
CARACAS 106. Tel. 32 23 01/33 26 04/33 24 73

YUGOSLAVIA – YOUGOSLAVIE
Jugoslovenska Knjiga, Terazije 27, P.O.B. 36, BEOGRAD.
Tel. 621-992

Les commandes provenant de pays où l'OCDE n'a pas encore désigné de dépositaire peuvent être adressées à :
OCDE, Bureau des Publications, 2 rue André-Pascal, 75775 PARIS CEDEX 16.
Orders and inquiries from countries where sales agents have not yet been appointed may be sent to:
OECD, Publications Office, 2 rue André-Pascal, 75775 PARIS CEDEX 16.

OECD PUBLICATIONS, 2 rue André-Pascal, 75775 Paris Cedex 16 - No. 41 375 1979
PRINTED IN FRANCE
(DH 92 79 02 1) ISBN 92-64-11936-1